Genes, Technology, and Apocalypse

Yochai Ataria

Genes, Technology, and Apocalypse

An Epilogue for Humanity

palgrave
macmillan

Yochai Ataria
Tel Hai Academic College
Upper Galilee, Israel

ISBN 978-3-031-59196-9   ISBN 978-3-031-59197-6 (eBook)
https://doi.org/10.1007/978-3-031-59197-6

This Palgrave Macmillan imprint is published by the registered company Springer Nature Switzerland AG.
The registered company address is: Gewerbestrasse 11, 6330 Cham, Switzerland

Paper in this product is recyclable.

This book is translated by Shmuel Sermoneta-Gertel.

We shall not cease from exploration
And the end of all our exploring
Will be to arrive where we started
And know the place for the first time.
Through the unknown, remembered gate
When the last of earth left to discover
Is that which was the beginning;
……
A condition of complete simplicity
(Costing not less than everything)
And all shall be well and
All manner of thing shall be well.
T. S. Eliot, *Four Quartets*

We children of the future, how could we be at home in this today? We feel disfavor for all ideals that might lead one to feel at home even in this fragile, broken time of transition.
Nietzsche, *The Gay Science*

That which is primally early shows itself only
ultimately to men.
Heidegger, *The Question Concerning Technology*

Acknowledgments

First and foremost, I would like to thank two people who have accompanied me throughout the process of writing this book: Lia Tamir and Dr. Amit Kravitz. Lia and Amit played a significant role in the years-long discussion at the heart of the present volume, and read and commented on numerous drafts. They have my sincere and profound gratitude.

Thank you to Prof. Adi Ophir, who read an early draft of the book and made a number of comments that led me to rewrite parts two and three and helped me understand levels of meaning that had escaped me even as I wrote. Thank you to Dr. Omer Horovitz, Prof. Ayelet Shavit, and Prof. Gonen Sharon, who read early versions of the book and helped me support my arguments.

I would like to thank all copyright owners for permission to use their images in this book.

Thank you to Yasmin Lev who prepared the book for publication. Thank you to David Shimonovich who, as he has done time and again for the past 20 years, edited the book and made it more readable.

I am also grateful to the publisher, Palgrave Macmillan, and its dedicated and professional team for their help and support.

I would like to thank Tel Hai Academic College for its generous financial support of the work on this book.

Contents

List of Figures

Introduction

Every finding, no matter how small or insignificant, that sheds some light on the origins of humankind, draws tremendous media attention. I will argue that this stems not only from a desire to confirm the theory of evolution, but also from our need to explain the horrors of the twentieth century. We long to discover the bones of the "first human"—club in hand, skull smashed in, lying next to another skeleton in the exact same condition. Such a discovery would allow us to claim, or so we would like to believe, that we were evil from the very beginning, that we alone survived, because we eliminated all of our rivals. The Neanderthals may have been better suited to the cold European climate, but we were natural born killers—hence the "great leap forward" of *Homo sapiens.* If proven, this theory has the power to absolve us of responsibility: "It is in our nature". When naturalist and Nobel laureate Konrad Lorenz focuses on aggression and the fighting instinct; when Richard Dawkins, one of the most influential biologists of the last four decades, chooses to focus on the selfish gene; when geographer and Pulitzer Prize-winning author Jared Diamond chooses to focus on the male chimpanzee; when world-renowned evolutionary psychologist Steven Pinker (named one of *Time*'s "100 Most Influential People" in 2004) chooses to focus on violence between very specific hunter-gatherer tribes—that is precisely what they are doing.

I will show that our need to be exculpated for the things we did during the course of the twentieth century (Fig. 3) causes us to systematically ignore important findings. For example, scientists have only recently begun to take a closer look at the theory that *Homo sapiens* carried diseases to which the Neanderthals had no resistance (Africa was a paradise for parasites, viruses, and bacteria). So too, we are constantly reminded of the similarity between humans and male chimpanzees, but rarely of the fact that we are just as closely related to bonobos—a far less violent species of chimpanzee that is, genetically speaking, just as similar to us as the common chimpanzee. It is not a coincidence. The idea that "human nature" resembles that of the male chimpanzee has the power to explain the killing fields of the twentieth century. More importantly, however, it gives us an alibi: "It's not us; it's in our genes!"

The quest for the bones of "the first human" is thus no less and perhaps even more a matter of moral than scientific interest—an explanation of our murderous behavior and, consequently, an acquittal. However, since unequivocal proof of the fact that we are "born killers" is hard to come by, we have a plan B: blame technology. For example, Albert Speer, one of the leading figures in the Nazi regime, concluded his final speech at the end of his Nuremberg trial with the following words: "There is nothing to stop unleashed technology and science from completing its work of destroying man which it has so terribly begun in this war" (Speer, 1970, p. 521). According to this approach, science and technology, rather than human beings, are responsible for the genocide of the Jews.

A good example of this can be found in the opening scene of Stanley Kubrick's 1964 film *2001: A Space Odyssey*, when an ape first learns to use a tool (a bone) and immediately sets about killing another living creature with it (Fig. 1). This scene encapsulates the way in which we think about the relationship between humans and technology and, as such, has become a founding myth of Western culture. It was only a matter of time before humans employed technology to take over the planet. It is technology that makes us who we are, and it is technology that made the *great leap forward* possible. Viewed in this light, technology possesses a dynamic of its own, and its development—from bones and stones to artificial intelligence—is a continuum.[1] Technological development is self-propelled and we are dragged along with it, while maintaining an illusion of control. The idea itself may be simplistic, but, once again, it allows us to relieve ourselves of responsibility: "It's not us; it's technology!"

I will show how these two vectors—the search for the bones of the first human and the dynamic of technology—operate within the same

[1] Later in the book, I will distinguish between technology, that is, technologies or tools, and **t**echnology, denoting an all-encompassing state of being.

Fig. 1 *2001: A Space Odyssey*

framework and towards the same goal: the absolution of humanity from responsibility for our actions. I will also show how, in a collective and apocalyptic state of mind that seeks both to return to the beginning of time and to precipitate its end, these two ostensibly opposing vectors converge: While the search for the bones of the first human is motivated by the desire to reveal the mysteries of the beginning of time, technology pushes us towards the end of time.

* * *

The burning question of our time is how much lower can we go? How far can the human race sink? Are Auschwitz and Hiroshima merely a taste of things to come? Bob Dylan provided an answer to this question in his final album of the twentieth century, *Time Out of Mind* ("It's not dark yet but it's getting there"), written as a summary of and reflection on a century characterized by technological-rational thinking and industrialized murder, a century in which "humanity has gone down the drain" ("Not Dark Yet").

It is hard to argue with facts. Hence, a completely different strategy is adopted: innocence on the basis of "human nature". We embarked on an evolutionary odyssey—seeking to discover the nature of the "first human" in order to understand the present. Make no mistake; the only possible path to exoneration lies in the determination that the first ancestor was inherently evil. Only such a finding can absolve us of responsibility. And if the first ancestor cannot be blamed, let us blame technology, for dragging us, *nolens volens*, to where we are today. Humankind is thus reduced to a hollow and indifferent species. When the goal is complete exoneration, sacrificing the ability to choose and control our actions and our fate is a small price to pay. Perhaps this is why current scientific discourse presents the problem of free will as nothing but an illusion.

Scientific curiosity ostensibly plays a role in our desire to discover the bones of the first human. On a deeper plane, however, this desire appears to stem from our need to explain the present and, thereby, perhaps have something to say about the future as well. In the words of Martin Heidegger—possibly the most important philosopher in the twentieth century and certainly the most controversial, due to his ties to the Nazi movement—in *The Origin of the Work of Art*: "A genuine beginning, as a leap, is always a head start, in which everything to come is already leaped over, even if as something still veiled. *The beginning already contains the end latent within itself*" (2008a, p. 201)[2]; or, in the words of the French philosopher Michel Foucault, one of the most influential thinkers of the final third of the twentieth century: "[The] starting-point is at the same time outside real time and inside it: it is the first fold that enables all historical events to take place" (2005, p. 359). Where *Homo sapiens* is concerned, it is a hopeless struggle:

> When he tries to define himself as a living being, he can uncover his own beginning only against the background of a life which itself began long before him ... Man ... is the being without origin, who has 'neither country nor date', whose birth is never accessible because it never took 'place'. (ibid., pp. 359–361)[3]

What Foucault is saying is that we will never be able understand humanity in terms of evolutionary continuity, no matter how many archaeological findings we unearth. That is because humanity is an idea that is redefined in every age in accordance with the spirit of the time. Our perception of humanity today is the result of Enlightenment thought: "Before the end of the eighteenth century, *man* [emphasis in the original] did not exist—any more than the potency of life, the fecundity of labour, or the

[2] Unless otherwise noted, all emphases, throughout the book, are mine.

[3] To correspond with the terminology used by the thinkers addressed in this text, the terms "man" and "humans" will be used interchangeably.

historical density of language. He is a quite recent creature, which the demiurge of knowledge fabricated with its own hands less than two hundred years ago" (ibid., p. 336). "Man" is thus an idea and no more.

This insight does not prevent us from seeking out the origin: the gene, the neuron, the atom. These are the basic units we hope will provide some explanation—perhaps more than an explanation, a justification. To take it a step further, following Foucault, one might say that the way in which we think about humanity after Auschwitz shapes evolutionary research. Humanity is an idea, and in the spirit of evolutionary scholarship, which has dominated discourse for decades, we will always seek to anchor our perceptions of humanity in natural terms.

The fact that *Homo sapiens* dominates the world compels us to discover what is unique about us. Ultimately, however, we fear discovering that the answer has already been given at Hiroshima and Auschwitz, which, in the words of Israeli philosopher Adi Ophir, have turned "from a private name into the family name of the victims of the West—the Native Americans, the Africans, the Japanese of Hiroshima, and all the rest" (2005, p. 550). We fear that what is unique about us may be our ability to commit genocide, to annihilate members of our own species (Fig. 3). We can go even further, however. Auschwitz, as a generic concept, is like the big bang—the gene that explains the difference between *Homo sapiens* and *Homo neanderthalensis*—the last *Homo* species to have coexisted with us, mysteriously disappearing some 35,000 years ago. The argument is that Auschwitz is not merely the family name that includes Hiroshima, the Gulag, Rwanda, and more, but the Archimedean point of Western culture, the uncanny origin: "It is no longer origin that gives rise to historicity; it is historicity that, in its very fabric, makes possible the necessity of an origin which must be both internal and foreign to it: like the virtual tip of a cone" (Foucault, 2005, p. 359). Auschwitz and Hiroshima are code names that signify not only the murder of God and the death of humanity, but also the moment at which the end and the beginning were fused—like the big bang and the black hole rolled into one. It is the zero point not only of Western culture but of humanity in general. On 6 August 1945, it was proven beyond a shadow of a doubt that the possibility that humanity will destroy itself, along with much of life on earth, is very real.

* * *

The point of departure of this book is that within Western culture there is a degree of resistance to the working-through of certain extreme traumas, which have consequently become foundational and even sacralized, like the Fall from Eden. In the words of historian and member of the American Academy of Arts and Sciences Dominick LaCapra, author of the seminal work *Writing History, Writing Trauma*: "There has been an important tendency in modern culture and thought to convert trauma into the occasion for sublimity … an entry into the extraordinary". Thus, extreme events, "such as the Holocaust or the dropping of atomic bombs on Hiroshima and Nagasaki, may become occasions of negative sublimity or displaced sacralization" and "may also give rise to what may be termed founding traumas—traumas that paradoxically become the valorized or intensely cathected basis of identity for an individual or a group" (2014, p. 23). This, in turn, leads to the creation of a new religion, just as the Holocaust became a "sacralized center of a civil religion" (28). When we are unable to relate to trauma as a historical event and conceive it as "constitutive of existence" (ibid., p. 49), it becomes a myth of origin: "the genesis of a myth wherein trauma is enacted in a story or narrative from which later traumas seem to derive" (ibid., p. 82). I would thus argue that Auschwitz and the bombing of Hiroshima have become the focus of a new civil religion—a founding trauma that has become what LaCapra defines as a myth of origin, and, as such, both the starting and end point of Western culture.

In the spirit of LaCapra's observations, I would add a caveat here: discourse that views Auschwitz not only as a synonym for other traumatic events but also as a code name for Western culture is obviously reductive and overly superficial. It goes without saying that Auschwitz, Kolyma, Hiroshima, Rwanda are distinct from one another, just the Muselmann and the *dokhodyaga* (the living corpse of the Gulag),[4] although physically similar, differ from one another in terms of historical circumstance and, no less importantly, in terms of subjective experience. Despite this reservation (of which we must never lose sight), in the context of our obsession with reduction, Auschwitz comes along and the mind of the first human becomes the mind of the last human: the first human, whoever that may be, and the last human, who appeared at Auschwitz, are mirror images of one another and it is unclear who came first, or, in the words of French philosopher Bernard Stiegler, "The end of the human cannot be investigated without investigating its origin, just as questioning death is questioning birth in a mirror" (1998, p. 135).

[4] For wider debate see: (Ataria, 2019, 2020).

Fig. 2 Auschwitz and Hiroshima—The Muselmann and the Hibakusha (被爆) We are convinced that if only we could decipher the true meaning of these code words, Auschwitz and Hiroshima, we would finally understand "human nature" *Right*: Yehuda Bacon (b. 1929), *Muselmann*, 1947, oil on cardboard, 32.1 × 48.6 cm; Yad Vashem Collection, Jerusalem. *Left*: Yasuko Yamagata, *Woman and Child Statue*, 1974

In any event, when Auschwitz and Hiroshima are seen as the end point (Fig. 2), all that remains is to chart the path to this point, as if it could not be otherwise. In this sense, it is not the first human who "caused" Auschwitz. Rather, it is Auschwitz that preceded and explains the first human. First came the Muselmann, and only then could *Homo sapiens* make the "great leap forward" from Africa to Europe.

* * *

French philosopher and sociologist Jean Baudrillard, a central figure in postmodern discourse, wrote, in *Simulacra and Simulation*: "There are cultures that have no imaginary except of their origin and have no imaginary of their end" (2004, p. 164), although it would seem that human culture in general and certainly white, Western, male, monotheistic culture is exactly the opposite: all we can think about (and thus from within) is the end. The theory of evolution, for example, sees *Homo sapiens* as no more (and no less) than an interim stage of development—yet, as Foucault suggests, "even though he [Man] is not conceived as the end-product of evolution, he is recognized to be one extremity of a long series" (2005, p. 341). From the moment the idea of man was born, it seems impossible to think of anything but its end: "As the archaeology of our thought easily shows, man is an invention of recent date. And one perhaps nearing its end" (p. 422). Common wisdom today is, as Ophir puts it, that "it may well be that we're now producing the tools and creating the conditions through which and due to which other people will bring an end to the world" (2005, p. 581). Thinking of the end, or rather, being-toward-the-end state of mind is, in effect, a precondition for our ability to think in terms of humanity as a whole. In other words, it is only through our common end that we are able to contemplate the concept of humanity not as individuals but as a single, homogeneous whole:

> The historicity of the present era is a historicity of an end, the moral call arising from humanity verging on the finish line. This end is a common one, and the presencing of the common finitude is the basis for a new, imposed universal association.... Finitude is not what separates but what lays common ground. (ibid., p. 621)

Ophir further stresses that living within an end state of mind has dramatic ramifications on daily practices: "In this era, the end of the world is present within the world, and every attempt at historical self-understanding necessitates anticipation of the end of the world". Essentially, the world not only "appeared within the world itself, and ... is present ... in vain eschatological beliefs but [also] in daily practices" (pp. 621–622).

It is the apocalypse that best defines our reality. Since the apocalypse is obsessed with the beginning and end of time, it comes as no surprise that Western, monotheistic culture is constantly in pursuit of the beginning, while seeking to precipitate that which lies at the end of a foreseen trajectory—that is, the desire for the end, the desire of the suicide.

From the moment the atomic bomb became possible it also became inevitable. The reason for this lies in the fact that we were thrown, a priori and in the deepest possible sense, into what sociologist Jacques Ellul called the "technological society" (Ellul, 1964)—a society that has completely adopted the internal logic of technology: "Anything that can be done will be done," and "If it can be done, it should be done." It is a reality of borrowed time, a fatalistic mood, a function of technological determinism that should be understood in the context of technological autonomy, which precludes the possibility of free will or, as German-Jewish philosopher and sociologist Herbert Marcuse put it, "In this universe, technology also provides the great rationalization of the unfreedom of man and demonstrates the 'technical' impossibility of being autonomous, of determining one's own life" (1968, p. 158).

It is convenient to explain reality in terms of an autonomous and deterministic technological system and to render those who create evil banal (it is not the evil that is banal but the way in which it is created). In this sense, German-Jewish philosopher Hannah Arendt (2006) hit on the most convenient approach after Auschwitz; an approach that requires the replacement of the term *Homo sapiens* with that of *Homo faber*—a creature capable of controlling tools, surroundings, and feelings, a kind of Eichmann: an official tasked with overseeing the murder of millions, morally indifferent, a technician and no more. The Jewish philosopher Gunther Anders, Arendt's first husband, who, like her, was greatly influenced by Heidegger, with whom Arendt had a romantic relationship when she was a student, expressed the idea as follows:

> The "technification" of our being: the fact that to-day it is possible that unknowingly and indirectly, like screws in a machine, we can be used in actions, the effects of which are beyond the horizon of our eyes and imagination, and of which, could we imagine them, we could not approve—this fact has changed the very foundations of our moral existence. Thus, we can become "guiltlessly guilty", a condition which had not existed in the technically less advanced times of our fathers. (Eatherly & Anders, 1962, p. 1)

I would argue that behind this idea lies the need to absolve ourselves of guilt. We are hollow; technicians for whom the world is technology and whose thinking is wholly technological. This explains the twentieth century, as well as the "great leap forward". In effect, we determine our nature on the basis of our commitment to technology as a way of thinking, of being. From here, it is but a short distance to Heidegger's shocking assertion (particularly in light of what we now know about his deep ties to Nazism) that "agriculture is now a motorized food industry, the same thing in its essence as the production of corpses in the gas chambers and the extermination camps" (Heidegger, 2012, p. 27). This is one side of the coin, also reflected in the words of the Jewish-American physicist Robert Oppenheimer, known as the "father of the atomic bomb": "The thing had to be done… [especially] at a time when all over the world men… were committed as never before… to technology as a way of life and thought" (Bird & Sherwin, 2005, p. 319).

As hard as it is to speak of Heidegger and Oppenheimer in the same breath—the former an anti-Semitic philosopher and Nazi who was not responsible directly for the murder of a single Jew, and the latter a Jewish physicist who led the American nuclear program (Project Manhattan) and was therefore responsible for the deaths of hundreds of thousands of Japanese—both absolve humanity of responsibility and lay the blame with technology. Annihilation is inevitable in the context of technological being.

The idea of commitment to a particular way of life is also reflected in research into the origins of humankind. Hollywood screenwriter turned anthropologist Robert Ardrey, who ingeniously grasped that there is no real boundary between Hollywood and scientific research, argued, in *The Hunting Hypothesis* (1976), that the transition from jungle to savannah was the watershed in human history:

> An evolutionary moment had to come for the evolving hominid when his life became more dependent on the successful kill than the secure escape … The renunciation of the trees meant a reversal of primate psychology: we could flee only so far and so fast; then we had to fight … We passed a point of no return in our commitment to the ancient, hazardous savannah. (Ardrey, 1976, p. 64–65)

The other side of the coin, the one that does not blame technology, goes something like this: it is human nature—like it or not, it is what it is. This is one of the reasons why we try to reduce humankind to the violent male chimpanzee. It is also the reason why we tend to describe the disappearance of the Neanderthals in terms of annihilation, despite the existence of other, non-violent explanations—for example, that the Neanderthals were unable to adapt to climate change, or that they interbred with *Homo sapiens*, meaning that they did not really go extinct, but became a part of us. Another possibility is that *Homo sapiens* may have been more successful at exploiting resources (some attribute this to differences in brain function), adversely affecting the Neanderthals without face-to-face violence. Some scholars have suggested that the Neanderthals' disappearance had nothing to do with *Homo sapiens*, but was the result of their limited gene pool.

Ignoring the bonobo in favor of the male chimpanzee, choosing to study particularly violent hunter-gatherer tribes or to interpret certain skeletal findings as reflecting battle injuries, and the idea that *Homo sapiens*

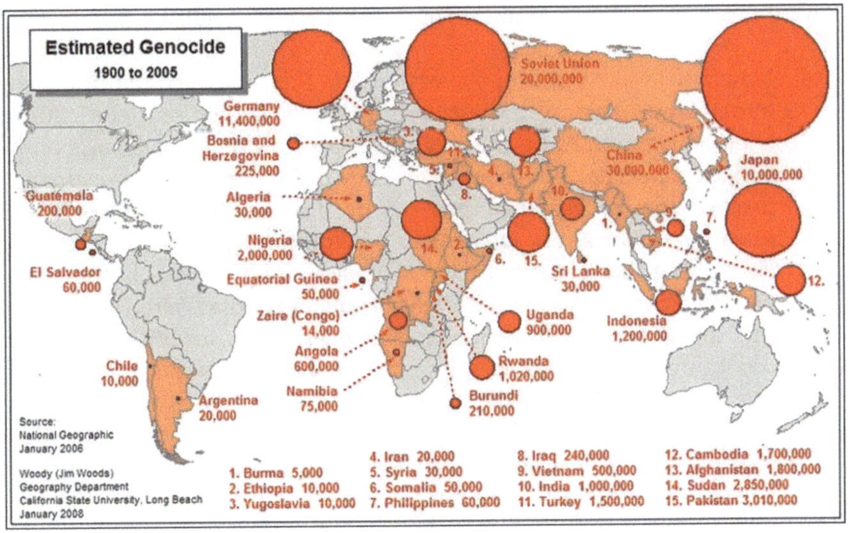

Fig. 3 Genocide in the Twentieth Century

conducted a "blitzkrieg" in Europe, thanks to their superior technology, wiping out the Neanderthals—all allow us to create a continuum between our "nature", as "demonstrated" by the findings, and our behavior in the present (Fig. 3). The various studies focus on violent genes (male chimps) and technological superiority (the elimination of the Neanderthals in Europe). These "findings" allow us to frame our behavior in the present. Both elements—the dynamic of technology and research into the origins of humankind—represent attempts to absolve humans of responsibility for the events of the twentieth century.

The present volume comprises three parts. Part I presents the first line of defense—*it's in our genes*; Part II presents the second line of defense—*it's all technology's fault*; while Part III links the vector that points towards the beginning and the vector that points towards the end to one another, demonstrating that they are, in fact, a single vector within the framework of *apocalyptic reality*.

* * *

Before moving on, I would like to add two methodological comments:

1. New articles are published every day on the subject of the origins of humankind, making it virtually impossible to represent the cutting edge of research. Although it is not the only field in which large numbers of papers are published, the study of human origins is especially problematic, because new findings do not merely reaffirm or refine previous theories, but also create new ones, often radically different from those that preceded them.[5] While writing this book, for example, the remains of Nesher Ramla *Homo* were discovered—a species that differs from both *Homo sapiens* and *Homo neanderthalensis* and, according to some scholars, fundamentally changes the way in which human history should be understood (Hershkovitz et al., 2021). Books that nevertheless insist on presenting the most up-to-date research quickly lose their relevance. Since the latest research is fascinating, and often contributes to the discussion at hand, I have chosen to reference some of it in the footnotes. Popular literature on the subject is a very different story. Questions pertaining to the origin of humankind have considerable resonance, transcending the narrow confines of academia. Time and again we find that it is not the discoveries themselves that are important, but the ways in which they are construed and rendered accessible to the public. In other words, it is not about the facts, but about the greater narrative into which they are woven. That is why I have chosen to explore popular science literature, on the question of human origin. I am interested in how popular culture assimilates them, what the public is told, and the unconscious archetypes on which the story rests, as well as its manifest and hidden goals.
2. In many ways, the book takes a Heideggerian perspective. Heidegger asks us to stop thinking of history as a linear timeline, a chronology of events and no more. History, according to Heidegger, exists within the present, organizing the world into which we are thrown. Historical eras are determined, that is, constituted, in accordance with the way in which we think of the world and act within it. Historical research, he argued, should focus on revealing the essential foundation that precedes historical events—a foundation that shapes and defines the situations within which we act. In *The Question*

[5] See e.g. (Dunsworth, 2010, p. 357).

Concerning Technology (1954), for example, Heidegger argues that although modern physical science began in the seventeenth century and machine technology developed only in the second half of the eighteenth century, the latter, in essential terms, did not follow the former. On the contrary, "modern technology, which for chronological reckoning is the later, is, from the point of view of the essence holding sway within it, the historically earlier" (Heidegger 1977, p. 22). Similarly, even if we do not believe in the Holy Trinity, as members of Western culture we are wholly governed by the Christian world view—particularly the book of Revelations, as we shall see later in the book. This is the approach I have taken in the present study and especially in the first part: considering the present in terms of the way in which it is shaped by history. At the same time, Heidegger is one of its central objects of criticism—explicitly in the second part and implicitly in the third.[6] It is my belief that such an approach, at this time, is not only possible but essential. In this, I have fully embraced Habermas's important assertion, made in 1953: "It appears to be time to think with Heidegger against Heidegger" (1998, p. 197).[7] It should be noted that adopting Habermas's position, like adopting the Heideggerian perspective in general, entails a kind of intellectual dualism, oscillating between criticizing Heidegger's philosophy and using it. At times, in this book, Heidegger will serve as the object of moral as well as philosophical

[6] In light of the nature of this project, it is important to note that when I speak of the *Heideggerian perspective*, I am referring to his later writings. Throughout the book and particularly in chapter two, I will ask whether a distinction can be made between an "earlier" and a "later" Heidegger, and if so, where the boundary should be drawn. The book *Being and Time* (1927), which many consider Heidegger's most important work, is undoubtedly part of his early thought. The question remains which books belong to his later thought. *The Question Concerning Technology* (based on a series of lectures delivered in Bremen in 1949 and published in 1954) is certainly in the latter category. Since this text stands at the heart of the present study, it is essential that we take a particularly critical approach to Heidegger's writings—all the more so when attempting to examine the characteristics of the current age through the lens of Heideggerian thought.

[7] In this context I would like to add the somewhat prophetic assertion that Heidegger made in a lecture course entitled "What Is Called Thinking" (1951–2): "We forget too easily that a thinker is more essentially effective where he is opposed than where he finds agreement." It is interesting to note that this was the first course he gave at the university following the five years in which he had been barred from teaching, as part of the denazification process. It was also his last significant lecture series.

criticism. At others, his writings will serve as the basis for discussion about technology and the type of thinking that constitutes the present era. In the analytical discourse, this type of conflict is generally viewed as a contradiction in need of immediate resolution. At this stage, however, we must have recourse to Heidegger's thought, which has shown that our ability to discern philosophy's most basic questions requires us to first act within it, discuss it, and think in its terms, and only then to overcome its limitations and its failings. We will thus act within Heidegger's philosophy, which appears, inevitably, to demand this back-and-forth motion, between involvement and detachment, affinity and rejection.[8]

* * *

In his book *Human, All Too Human*, in a section entitled "Inherited Faults of Philosophers", Nietzsche accuses philosophers of having failed in their attempts to study and understand man:

> All philosophers have the common fault that they start from man in his present state and hope to attain their end by an analysis of him. Unconsciously they look upon "man" as an *æterna veritas*, as a thing unchangeable in all commotion, as a sure measure of things. Everything the philosopher has declared about man is, however, at bottom no more than a testimony as to the man of a *very limited* period of time. Lack of historical sense is the family failing of all philosophers; many, without being aware of it, even take the most recent manifestation of man, such as has arisen under the impress of certain religions, even certain political events, as the fixed form from which one has to start out. They will not learn that man has become, that the faculty of cognition has become; while some of them would have it that the whole world is spun out of this faculty of cognition. Now, everything *essential* in the development of mankind took place in primeval times, long before the four thousand years we more or less know about; during these years mankind may well not have altered very much. But the philosopher here sees "instincts" in man as he now is and assumes that these belong to the unalterable facts of mankind and to that extent could provide a key to the understanding of the world in general: the whole of teleology is constructed by speaking of the man of the last four millennia as of an *eternal* man towards whom all things in the world have had a natural relationship

[8] In this passage, as in others pertaining to Heidegger, I have benefitted greatly from the help of Lia Tamir, for which I am grateful.

> from the time he began. But everything has become: there are *no eternal facts*, just as there are no absolute truths. Consequently what is needed from now on is *historical philosophizing*, and with it the virtue of modesty. (1996, pp. 12–13 [Emphases in the original])

Although Nietzsche wrote this text in the nineteenth century, philosophers and scholars continue to treat humans as if they were the finished product and then search for the starting point that would necessarily have led to humans as we perceive them today. Sigmund Freud, the father of psychoanalysis, offers a striking example of this approach. In *Totem and Taboo* (1913), Freud looks at the Christian Eucharist in its contemporary form and links it to the killing of the father in primitive societies. He thus approaches human evolution and mythology from a very specific perspective—that of a particular historical phenomenon he happened to be studying. He explains the present in terms of unconscious memories of events that occurred at the dawn of human culture—particularly the killing of the father. The evidence he relied on, however, was neither archaeological nor genetic, but rather his own observations regarding traumas encountered in the present—women who had suffered sexual abuse, and shell-shocked men in the wake of the First World War. These phenomena led him to claim that Western culture emerged from foundational trauma: the killing of the father, the Fall from Eden, the fear of castration anxiety, etc. Once again, however, Freud lacked concrete evidence, merely projecting current circumstances on the past and constructing a narrative around it without any real basis. Freud is but one example of the many scholars guilty of this fundamental sin. When Diamond compares the advantage of *Homo sapiens* over rival species in terms of a propensity for technological innovation, he too projects contemporary experience of ever greater and faster technological development on the distant past.

In Western culture, this sin—relating to the present as if it were the only possible outcome, giving rise to historical, archaeological, and biological studies that necessarily lead to the current reality—seems practically inevitable. Even Nietzsche himself, in this very passage quoted above, falls into the same trap. Despite his claim that "there are *no eternal facts*, just as there are no absolute truths", he still relies on the evolutionary approach, whereby the starting point determines the end point: "Now, everything *essential in the development of mankind* took place in primeval times, long before the four thousand years we more or less know about; during these years mankind may well not have altered very much." Although Nietzsche attempts to look to (what he perceives as) the distant past, he too believes that the evolutionary process occurred at the dawn of humankind and determined who we are today. In this sense, he is no different from the philosophers he criticizes.

Nietzsche accuses philosophers of having failed to look upon man as some-*thing* (not an entity or some-*one*) fluid, changing, and dynamic. This accusation is all the more apt when it comes to those who treat science from a perspective that I will define, in a somewhat simplified fashion, as materialist-evolutionary—of which Diamond, Pinker, Harari, and Dawkins are prime examples in popular discourse, representing the mainstream. Any study that seeks an anchor or starting point accepts the premise that such an anchor determines something essential about who we are today, leaving only the question of which starting point to choose.

This approach, apparently typical of Western culture, is expressed as follows in T. S. Eliot's "Little Gidding":

> We shall not cease from exploration
> And the end of all our exploring
> Will be to arrive where we started
> And know the place for the first time.
> Through the unknown, remembered gate
> When the last of earth left to discover
> Is that which was the beginning;
> ……
> A condition of complete simplicity
> (Costing not less than everything)
> And all shall be well and
> All manner of thing shall be well.
> (Eliot, 1972, p. 59)

Alongside this approach, which seeks to understand the beginning, we find another, equally strong tendency: the belief that there is something at the absolute end that allows us to understand, perhaps for the first time, what happened at the starting point—or, in the words of Heidegger: "That which is primally early shows itself only ultimately to men" (1977, p. 22). When we read this sentence, we discover that although Heidegger does reject both scientific determinism and biologism, when it comes to the technological essence, he does not hesitate (certainly in his later period) to adopt a kind of technological fatalism.

* * *

A prominent example of this addiction to ends and beginnings is Israeli historian Yuval Noah Harari, whose books on the history of humankind and its future (2014, 2016) have enjoyed tremendous success in popular discourse in recent years. In these books, Harari proposes several theses, principally that: (i) humans are the deadliest species on the planet; (ii) the techno-scientific project is about to defeat old age and death (Gilgamesh project); (iii) history is approaching the singular moment in which humans will upgrade and recreate themselves as Homo Deus (the superhuman). The reason for the success of Harari's books lies not only in their ability to give readers the (illusory) feeling that they have gained deep insights into human history,[9] but also in the way that they embrace our most basic yet hidden assumptions about humankind: (1) man is a cruel creature with no self-control or self-restraint—making humans the most dangerous species on earth, (not only) to other species but also to themselves; (2) the techno-scientific project has a clear plan of action—the superhuman (the amalgamation of theses ii and iii).

Harari creates a temporally and spatially coherent narrative that successfully upholds these two principles: the murderous human on the one hand, and technology as an autonomous system that careers towards the singular point on the other. Consciously or unconsciously, these two principles negate human freedom, making us slaves to our genes and to technology. These principles lend themselves to two narratives that move in two seemingly opposite temporal directions—the vector that goes back in time to the bones of the first human, to the migration from Africa to Europe, to the male chimpanzee, and the vector that goes forward in time to the superhuman. These narratives are based on two corresponding principles: the murderous human on the one hand and the technological-singular dynamic on the other. While we search for and identify the murderous human in the evolutionary-deterministic vector, we seek the superhuman in the technological-fatalistic dynamic.

[9] The impression is false, however, because it is precisely the complete lack of innovation and originality in Harari's books that has made them so successful.

The covers of Harari's first and second books in their Hebrew editions leave little doubt on this matter. The first cover links cave paintings, which many feel symbolize our ability to represent the world and stand at the basis of the great leap forward (a singular point at which *Homo sapiens* became the dominant species) and the moon landing—another singular point on the path of humanity's departure from Earth. The second cover features the next singular point—the transformation of *sapiens* into a cyborg of sorts, a superhuman technological being (a precondition for our departure from Earth). Both covers focus on singular points, and indeed the writing of Yuval Noah Harari, like that of other secular prophets—Nietzsche, Marx, Kubrick, Kurzweil, and others—revolves around the singular point. The problem is that thinking in terms of the singular point is end-times thought, thoroughly steeped in the apocalyptic (Fig. 4).

Fig. 4 The longing for ends and beginnings
Our motivation to find the bones of the first human is not merely scientific (in the narrow sense of the term). Since we are steeped in the apocalyptic state of mind, we are convinced that the bones of the first human will also tell us something about our end. Indeed, in apocalyptic thinking, there is an inherent link between beginning and end

It is worth recalling, at this point, Nietzsche's well-known formulation in *The Gay Science* (1882):

> After Buddha was dead, his shadow was still shown for centuries in a cave—a tremendous, gruesome shadow. God is dead; but given the way of men, there may still be caves for thousands of years in which his shadow will be shown. —And we—we still have to vanquish his shadow. (1974, p. 167)

Harari is only a case in point. The two principles—the murderous human and the singular technological dynamic—common to many contemporary thinkers, stem from a far more elementary notion. The discourse surrounding evolution and technology is imbued with apocalyptic thought: the desire to be saved from current reality and to reach the end times. It is worth stressing that those who accept the apocalyptic approach (common to some currents in Judaism, Christianity, and Islam), according to which time is linear and has a beginning and end, believe that history has meaning, direction, and an ultimate goal. Those who exist within the apocalyptic state of mind are driven by the obsession to expose the beginning and a desire to reach the inevitable end. The French philosopher Jean-Luc Nancy (2017), for example, wrote that Heidegger's longing for a "new beginning" for Western culture, led him to suggest a deep philosophical justification for the annihilation of the Jewish people—one that resulted, according to Heidegger, from Judaism's a priori tendency to self-destruction. The desire to expose the beginning stems, as noted, from the belief that the starting point reveals a fundamental truth about the end point. The story of the murderous ape and the hunter-gatherer is therefore the story of the atomic bomb, the story of the Neanderthals' disappearance, or as Lacoue-Labarthe wrote, 'In the Auschwitz apocalypse, it was nothing less than the West, in its essence, that revealed itself—and that continues, ever since to reveal itself' (quoted in Vioulac, 2021, p. 66).

Western thought cannot break free, it would seem, from the idea of a unified and linear arrow of time progressing from beginning to end, and even if that end is as yet unknown, when it does arrive, we will perceive everything as a single unit. The goal of this present volume is to reveal this metaphysics and how it shapes philosophical, scientific, and cultural discourse, and ultimately, how it defines the way in which we are thrown into the world.

* * *

Disclaimer and Warning: I would like to warn readers: the present author is also a product of Western culture and is therefore equally incapable of escaping this metaphysics and this obsession with beginnings and ends. Since I cannot write outside myself, I suppose I cannot avoid falling into these pitfalls either, but if Nietzsche stumbled here, I will allow myself to do the same.

References

Ardrey, R. (1976). The hunting hypothesis: A personal conclusion concerning the evolutionary nature of man. Atheneum.

Arendt, H. (2006). Eichmann in Jerusalem: A Report on the banality of evil. Penguin Books.

Ataria, Y. (2019). Becoming nonhuman: The case study of the Gulag. Genealogy, 3(2), 1–27. https://doi.org/10.3390/genealogy3020027

Ataria, Y. (2020). Investigating the origins of body-disownership: The case study of the Gulag. Journal of Phenomenological Psychology, 51(1), 44–82.

Baudrillard, J. (2004). Simulacra and simulation (S. F. Glaser, Trans.). University of Michigan Press.

Bird, K., & Sherwin, M. J. (2005). American Prometheus: The triumph and tragedy of J. Robert Oppenheimer. Atlantic Books.

Eatherly, C., & Anders, G. (1962). Burning conscience: The case of the Hiroshima pilot, Claude Eatherly, told in his letters to Günther Anders. Monthly Review Press.

Eliot, T. S. (1972). Four quartets. Faber and Faber.

Ellul, J. (1964). The technological society (J. Wilkinson, Trans.). Vintage Books.

Foucault, M. (2005). The order of things: An archaeology of the human sciences. Routledge.

Habermas, J. (1998). Martin Heidegger: On the publication of lectures from the year 1935. In R. Wolin (Ed.), The Heidegger controversy: A critical reader (pp. 186–197). MIT Press.

Harari, Y. N. (2014). Sapiens: A brief history of humankind. Vintage Books.

Harari, Y. N. (2016). Homo Deus. Random House.

Heidegger, M. (1977). The question concerning technology and other essays (W. Lovitt, Trans.). Harper & Row.

Heidegger, M. (2008a). The origin of the work of art. In M. Heidegger & D. F. Krell (Eds.), Basic writings: From being and time (1927) to the task of thinking (1964) (pp. 139–212). Harper Perennial.

Hershkovitz, I., May, H., Sarig, R., et al. (2021). A middle Pleistocene homo from Nesher Ramla, Israel. Science, 372(6549), 1424–1428. https://doi.org/10.1126/science.abh3169

LaCapra, D. (2014). Writing history, writing trauma. Johns Hopkins University Press.
Marcuse, H. (1968). One dimensional man: Studies in the ideology of advanced industrial society. Beacon Press.
Nancy, J. L. (2017). The banality of heidegger (J. Fort, Trans.). Fordham University Press.
Nietzsche, F. (1974). The gay science (W. Kaufmann, Trans.). Vintage Books.
Nietzsche, F. (1996). Human, all too human. Cambridge University Press.
Ophir, A. (2005). The Order of evils: Toward an ontology of morals (R. Mazali, & H. Carel, Trans.). Zone Books.
Speer, A. (1970). Inside the third Reich (R. Winston, & C. Winston, Trans.). Weidenfeld and Nicolson.
Stiegler, B. (1998). Technics and time, 1: The fault of Epimetheus. Stanford University Press.
Vioulac, J. (2021). Apocalypse of truth: Heideggerian meditations (M. J. Peterson, Trans.). The University of Chicago Press.

PART I

It's in Our Genes

The story of the human race is War. Except for brief and precarious interludes there has never been peace in the world; and before history began murderous strife was universal and unending.
Winston Churchill, 1929

Judging by the sheer quantity of academic and popular material on the subject, the question of humanity's origins seems to have gone from a matter of scientific enquiry to a veritable obsession in the second half of the twentieth century. This stems not only from a desire to substantiate the theory of evolution, which Israeli-French philosopher Daniel Milo has called "the theory closest to definitive proof one can hope for in science" (2009, p. 176), but from the need to explain the history of humankind.[1]

We strive to explain how the most enlightened, educated, and advanced people on Earth built Auschwitz, dropped atomic bombs, and sent millions to the Gulag—three events that occurred in the twentieth century and defiled the human race.

In Part I, I will argue that much of our interest in human history in general and in the development of the genus *Homo* in particular (from the

[1] Note that the original title of Milo's book was *The Invention of Research* (*Hamtza'at hameḥkar*), not *The Invention of Tomorrow* (*Hamtza'at hamaḥar*). Unless otherwise noted, all translations from non-English sources are by the translator of the present volume.

chimpanzee, through *Australopithecus*, all the way to *Homo habilis*, *erectus*, and *sapiens*) derives from the desire to show that we are inherently defiled and thereby obtain technical exoneration from the horror.

Since the facts themselves are indisputable (see Fig. 3), we can only plead temporary insanity of the human race during the course of those events. The further we delve into human history, however, the more we realize that temporary insanity is not an option. Insanity—maybe; temporary—not a chance. Genocide is deeply embedded in human culture. For example, after the Israelites massacre all of the male Midianites, Moses commands them to finish the job: "And now, kill every male among the little ones, and every woman who has known a man in lying with a male, kill. And all the little ones of the women who have not known lying with a male, let live" (Num. 31:17–18; trans. Alter). And Moses is not alone. On 24 August 1572, during the French Wars of Religion, between Catholics and Huguenot Protestants, a 22-year-old King Charles IX shouted, perhaps at a moment of nervous breakdown: "Kill them all! Let no one escape to upbraid me!" On that same day, thousands of Huguenots were slaughtered, in what came to be known as the St Bartholomew's Day massacre. That same desire to utterly annihilate the other is also reflected in the words of Adolf Eichmann, in a conversation with a Dutch former SS officer by the name of Willem Sassen: "I have to tell you quite honestly, that if of the 10.3 million… identified, as we now know, we had killed 10.3 million, I would be satisfied, and would say, good, we have destroyed an enemy" (Stangneth, 2016, p. 303–304).

We are thus left with only one possible explanation: "human nature". As Ardrey wrote in *The Hunting Hypothesis*: "*Man is man, and not a chimpanzee, because for millions upon millions of evolving years we killed for a living*" (1976, p. 17 [Emphasis in the original]). In the 1960s, Ardrey—one of the protagonists of this first part of the book—decided to leave his job as a Broadway playwright and Hollywood screenwriter, to write popular science. In the 1960s and 1970s, his books were a phenomenal success and had an immense impact on popular discourse.[2] Despite their simplicity (or perhaps because of it), Ardrey's books tell a deep truth about the chauvinistic-scientific "human nature" approach. As we will see in this section, not a lot has changed since Ardrey's books were first published.

[2] In 1969, *Time* magazine named Ardrey's *African Genesis* the most notable non-fiction book of the decade.

Proving that human nature is inherently, categorically, and undeniably evil justifies or at least explains the horrors of the twentieth century. Further evidence that this is indeed one of the goals of the study of human origins can be found in the works of one of the foremost authorities on the subject in the last third of the twentieth century, Richard Leakey, son of paleoanthropologists Louis and Mary Leakey. Leakey grew up in Kenya and worked on excavations with his parents. In his book *The Origin of Humankind*, he writes, apparently critically, that "the notion of 'Man the Killer Ape'... gained wide adherence, possibly because it seemed to explain (*or even excuse*) the horrible events of the war" (1994, p. 10). Of course, if it is simply "human nature", the concept of "evil" loses its power. Nature cannot be good or evil; nature is what it is—beyond good and evil. As I will attempt to show in this part of the book, this is, in fact, the ultimate goal of scientific research today: to reduce everything to nature, in an attempt to blame the smallest unit, the gene itself, the "root of evil", in order to obtain an acquittal. Hence the obsession.

We long to find the remains of the first human and pray that they will be found alongside those of another person, clubs in hand and skulls smashed in, in order to prove that, in the words of biological anthropologist Michael Ghiglieri, "Wars are older than humanity itself" (1999, p. 161). Should such a discovery be made, we will be able to plead innocence on the grounds of human nature—although the idea of "innocence" would be meaningless, because guilt—the driving principle in the Judaeo-Christian world—will have been reduced to nature. The primary goal of human evolutionary research is thus to exonerate us for what has been and perhaps for what is yet to be.

In *People of the Lake* (1979, pp. 28–33), Leakey and Lewin suggest that the search for our earliest ancestors (see fig. 1) also promises to provide the key to our long-term survival as a species. To this end, we must find the answers to questions like "Are humans innately aggressive?" Leakey and Lewin concede that, "*Homo sapiens* is the only species with an apparently well-developed proclivity for killing its own kind", yet they are reluctant to exonerate us, and attempt to prove that the phenomenon is not purely biological: "War is a cultural invention. It is not a biological inevitability." Consequently, "The eighty million victims of the two world wars died not as a result of a deep ineradicable urge that drove them and their leaders into combat, but rather as a consequence of political machination that brought nations into confrontation, something that humans alone can accomplish." Note that they blame the system while exonerating the

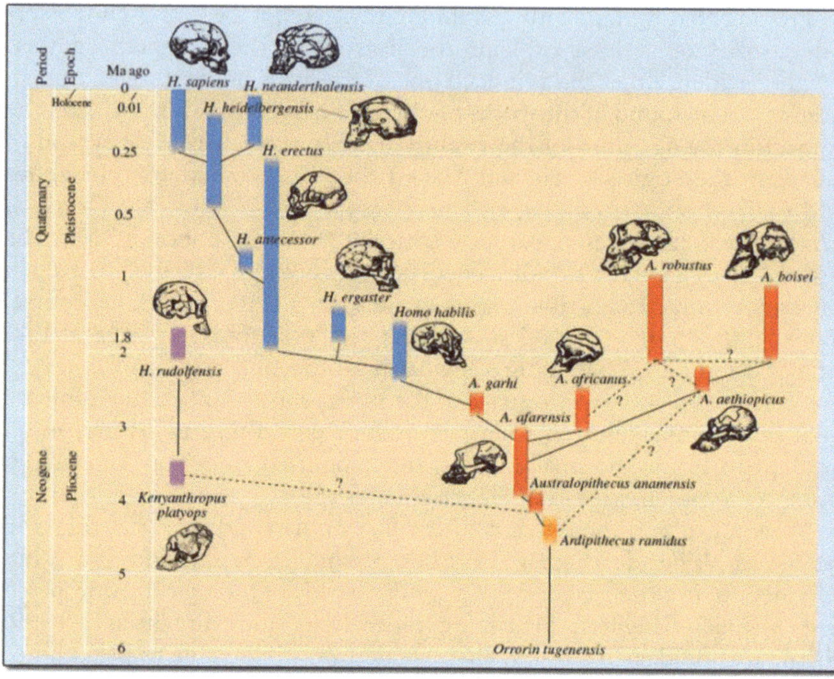

Fig. 1 Human evolution timeline (tentative)

individual. Despite their reservations, Leakey and Lewin ultimately conclude that humankind is defined by its ability to create the means of its own destruction. What makes us unique in the animal kingdom, the dominant species, is human culture.

This first part of the book addresses one of the central and most difficult questions in scientific and popular literature today: *how is it that we survived when all other Homo species disappeared?* In the first chapter, I examine various characteristics (such as brain size, representational ability, culture, language, and suspicion) in search of a possible explanation. The more scholarship advances, however, the blurrier the boundary between *Homo sapiens* and other species becomes. Our claim to uniqueness is all the more tenuous when even language is no longer a bastion. Could we alone have survived because we are inherently evil? In the second chapter, I explore the hunter-gatherer theory, which asserts that we are, first and foremost, hunters. Although this theory explains many phenomena on individual, familial, and communal levels, there is no apparent reason to prefer it to the gatherer-hunter theory, which posits that we are, above all, gatherers, and only occasional hunters. While according the first approach, humans are above all hunters, so that our evolutionary development—what made us us—is linked to hunting and blood and a very specific kind of masculinity, the second approach posits that pre-industrial humans were more similar to "noble savages", as defined by influential Enlightenment philosopher Jean-Jacques Rousseau (1712–1788), in "The State of War": "In the state of nature there is thus no general war of every man against every man; the human race was not created simply to destroy itself" (Roosevelt, 1987, p. 236).

The only advantage offered by the first theory is that it explains the horrors of the twentieth century and offers the possibility of exoneration: if we are inherently evil, we can hardly be blamed for the mass murders committed throughout human history. A similar approach can be observed in research on our closest genetic relative, the chimpanzee. There would appear to be a tendency to compare us to the violent male chimpanzee and to ignore another genetically similar cousin—the bonobo, a (relatively) non-violent pygmy chimpanzee. The tendency of scholars such as biologist and geographer Jared Diamond, one of the world's most influential popular science writers, to ignore the bonobo attests to the fact that the study of human origins seeks to demonstrate the existence of a single trajectory, leading directly from the violent male chimpanzee to the concentration and death camps. After discussing the killer ape theory in chapter three, chapter four asks what happened to the Neanderthal, why a species better adapted to the cold European climate died out some 35,000 years ago. Here too, the prevailing scientific theory (not necessarily supported by scientific evidence) tends to explain the disappearance of the Neanderthals as the result of *Homo sapiens'* first genocide—perceived not as an accident, but as a testimony to the unique nature of our species. As in the case of chimpanzee research, there are many other possibilities that fail to receive sufficient attention—from interbreeding to disease to gradual displacement, due to more successful hunting and gathering strategies. The section concludes with a discussion of humanity's first line of defense: don't blame us; *it's in our genes.*

CHAPTER 1

Modern Humans: Attempting to Draw a Dividing Line

There is probably no other issue more important to the general public than the evidence for the first appearance … of our own genus. We are special, we are different, and we are concerned. We face a very complicated future … and if we can really pin down this issue of who we are and where we are from once and for all, it would be to everyone's greater good.

R. E. Leakey, "Early Humans: Of Whom Do We Speak?"

The separation of humans from chimpanzees began some 6 million years ago, as the former began to acquire erect bipedal locomotion—the reasons for which remain unclear to this day but are probably related to climatic change and a reduction of forests, in favor of savannah environments.[1] There is evidence—in the form of fossilized skeletal remains, nicknamed Ardi (*Ardipithecus ramidus*)—that hominins walked erect (judging by pelvic structure) as early as 4.4 million years ago. Note that the appearance of erect locomotion does not, in itself, indicate the abandonment of

[1] Some studies reject this theory, suggesting that bipedalism was in fact an arboreal adaptation, facilitating safer and more efficient locomotion between branches (Drummond-Clarke et al., 2022). Similar reservations can be expressed regarding nearly every assertion in this section. In light of the nature of the project at hand, I have necessarily simplified complex ideas. Where I feel that a more extensive and nuanced approach is called for, I will provide references to the relevant literature in the notes.

Y. Ataria, *Genes, Technology, and Apocalypse*,
https://doi.org/10.1007/978-3-031-59197-6_1

arboreality—witness the long arms and fingers (especially suited to grasping branches), grasping toes, and bone structure of various apes. The development of erect locomotion not only included changes in the pelvis, feet, and other parts of the body, but also freed the hands, allowing them to perform a variety of tasks. While the process ended around 4 million years ago, two further significant "leaps" in human development occurred: increased brain volume and tool use.[2]

Contrary to previous assumptions, based on the palaeoanthropological studies conducted by Phillip Tobias in the 1980s, the significant change in the size and structure of the brain occurred around 1.8 (not 2.5) million years ago.[3] In *Homo erectus*, the change occurred between 1.5–2 million years ago and 50,000 years ago. It was once thought that the change in the hominin brain preceded the more sophisticated use of tools. More recent findings have shown, however, that basic tools were in use some 3.3 million years ago—before the appearance of *Homo*.[4] 800,000 years later (2.5 million years ago), we find a far more complex system of production and use of tools in a species considered the ancestor of the genus *Homo*: *Homo habilis* (1.65–2.3 million years ago)—a species identified with Oldowan culture (there is evidence of Acheulean culture from 2.6 million years ago).[5] The fact that *Homo habilis* is defined in terms of the tools they used attests to the importance of tools for their survival. From *Homo habilis*, through *Homo erectus*, to *Homo sapiens*, we see the growing importance of tools.

The transformation from ancient to modern human passed through four critical junctures: bipedalism, brain enlargement, control of fire, and tool usage.[6] In any event, half a million and even a hundred thousand years ago, there were relatively few hominids in general and *Homo sapiens* in particular in the world. Moreover, even when it comes to

[2] The word "leap" is used merely as a figure of speech here. In reality, evolutionary processes occur over long periods of time. This is equally true of references to the "appearance" of species and so forth.

[3] For an extensive discussion, see (Ponce de León et al., 2021).

[4] For an extensive discussion, see (Harmand et al., 2015).

[5] All of the numbers cited here are based on generally accepted estimates. There are, of course, differences of opinion among scholars. See (Braun et al., 2019).

[6] A distinction should be made between the use of naturally created fire (a million and a half years ago) and the ability to ignite and control it. Evidence of human control of fire some 800,000 years ago was discovered at Gesher Benot Ya'aqov, Israel. Some have argued (Wrangham, 2009) that the use of fire and cooking actually facilitated brain growth, allowing *Homo habilis* to evolve into *Homo erectus*.

60,000–70,000 years ago, we cannot yet identify, with any degree of certainty, the "great leap" that took *Homo sapiens* out of Africa and made them rulers of the world. Put simply, now that we alone have survived and all of our rivals are extinct, that is, from the perspective of the end result, we try to identify some relative advantage to which this might be attributed, but it is all hindsight. Were we to examine the situation at various points in the last 2 million years, we would find no particular advantage in *Homo habilis* or *Homo erectus* or *Homo sapiens* marking it as a species destined to change the reality of all living things on Earth. At best, we would see a miserable biped barely able to make it to the end of the day. Even something like the use of fire, which seems like such a dramatic development to us today, only assumes this significance with hindsight. Thus, even at a distance of half a million years, there is no apparent reason for the superiority of *Homo* over other species. What is it then that makes *Homo sapiens* so extraordinary? It is my contention in this book that we cannot but consider this question from within the context of the bodies burnt at Auschwitz and Hiroshima—the founding trauma of the age, a myth of origin, a story that defines the way in which we are thrown into reality and understand ourselves and the world.

Let us now return to what we do know. *Homo sapiens* (*sapiens*)—indistinguishable in appearance and cognitive ability from humans today—came to dominate Europe about 35,000 years ago. The question is, what happened to *Homo sapiens* during the period beginning some 300,000 years ago that unforeseeably led our species to dominate the world? In his book *The Invention of Tomorrow*, Milo stresses that "everything humans are capable of—tools, music, love—can be accomplished by at least one more animal, albeit in a less extensive and sophisticated manner", yet "no one truly believes that *Homo sapiens* is just one of thirty million or so surviving species". Milo argues that, "humans are the ultimate freaks, since they are exceptional in nearly every significant way" (2009, pp. 11–53). *Homo sapiens* is an "exaggerated creature". For example, our brains are wildly disproportionate to the size of our bodies, measuring 1400 cm^3, as opposed to the 400 cm^3 one would have expected in a hominid of our size (Fig. 1.1). Even our numbers are out of all proportion. The expected population size of an animal like *Homo sapiens* is less than half a million, yet there are nearly 8 billion of us!

Another area in which our exceptionality stands out is the high dependency of human offspring on caregivers, which would appear to be a real problem from an evolutionary perspective. In all dependent species, the

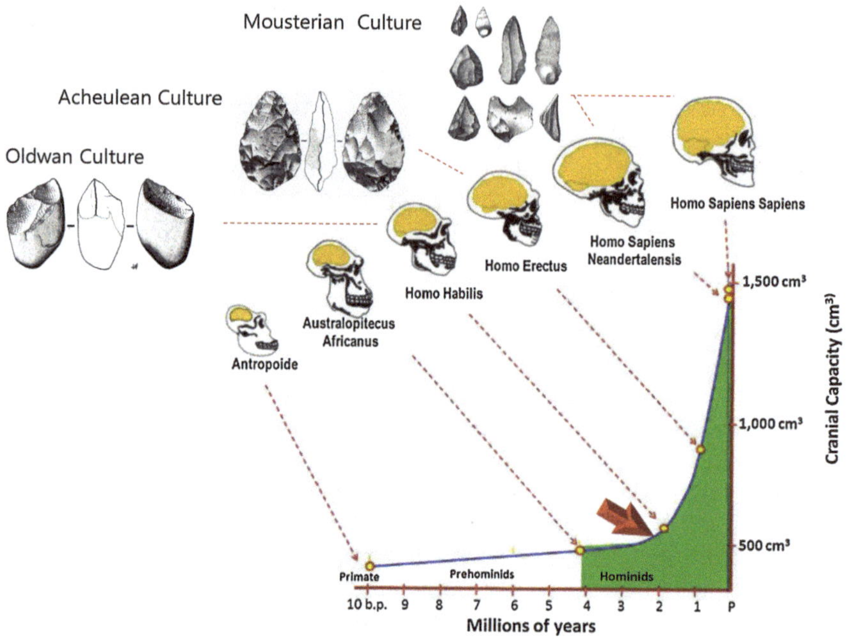

Fig. 1.1 Development of the human brain
On average, the primate brain is more than double the volume one would expect in a mammal of its size. Over the course of nearly 7 million years, the human brain trebled in size, with most of the enlargement occurring in the last 2 million. Precise determinations in time are elusive, since we have no ancient brains to compare to. What we can measure is ancient skulls. According to these measurements, for the first two-thirds of our history, our ancestors' brains were within the range of those of modern apes. Lucy's (*Australopithecus Africanus*; 2.9–3.6 million years ago) skull could probably contain a brain with a volume of 400–550 cm^3, compared to 400 cm^3 for a modern chimpanzee skull and 500–700 cm^3 for that of a gorilla. Of course, we have no way of knowing its connectivity density. During this period of time, members of Lucy's species began to present subtle changes in structure and form with respect to apes. For example, the neocortex began to expand, to reorganize its functions, from the elaboration of visual information to other regions of the brain. In the final third of the period, the brain nearly reached its final size. Homo habilis shows a very small increase in brain size, including an expansion of the part of the forebrain associated with language (Broca's area). The earliest known fossil of *Homo erectus*, from some 1.8 million years ago, had a brain size of slightly over 600 cm^3. The brain continued to grow slowly over the course of the

fetus is forced "to complete a significant part of its development outside the womb, and thus enters the world 'half-baked'". Once again, however, *Homo sapiens*, takes things to another level and is "unique among dependent species" (p. 116). To complete its development, the human brain would require 21 months of gestation, but the "pelvis and birth canal are suited to accommodate the head of a seventh-month-old foetus". Nine months are thus a compromise, resulting in the fact that "the human baby must enter the world with only about a quarter of the final volume of its brain" (p. 122). For the brain to continue to develop, it requires nearly three quarters of the daily calories consumed by a newborn.

In the discussion of the rise of *Homo sapiens*, cave paintings such as those of Lascaux, in south-western France (Fig. 1.2), and others are generally given considerable weight, because scholars believe they attest to a capacity for abstract thinking (geometric shapes and symbols that do not represent actual objects), a ritual framework of some sort, and cognitive and cultural development in general, possibly symbolizing modern humanity's leap forward. Or perhaps they represent the beginning of the ability of painters to reflect their inner worlds in their physical environment, rendering their subjective experience collective. These symbols may also indicate attempts to shape and control nature, or, as Richard Leakey and Roger Lewin put it

---

Fig. 1.1 (continued) next 500,000 years, reaching a volume of 1000 cm^3 or more. As culture and language became more complex, diet improved, and technology developed, our brains grew further to adapt to these changes (resulting, in turn, in more frequent changes, commensurate with brain size). Note that this assertion is somewhat simplistic, since we know of long periods in which the human brain continued to grow, even considerably, without any significant technological advances: "For more than a million years, that is, 80 per cent of the species' time on Earth, its magnificent brain never bothered to upgrade the tools it produced." For example, we know that technology remained exactly the same for 780,000 years, despite the fact that "Erect man's brain grew by 40 per cent during that period (from 850 to 1,100 cm^3)." Furthermore the (threefold) leap forward, in the form of "stone tools, controlling fire, and creating spears, occurred before the hominid brain had entered a positive feedback look" (Milo, 2009, pp. 79–85). These findings teach us that we cannot presume a one-to-one link between technology, culture, and biology
Image source: (Bretas et al., 2020)

Fig. 1.2 The Lascaux cave paintings
Lascaux is a series of interconnected caves, famous for its Upper Paleolithic wall paintings. The paintings primarily depict ungulates and large predators, most of which are known from fossil remains in the area. It is estimated that the paintings were created 17,300 years ago, probably for ritual purposes
The paintings in the Chauvet-Pont-d'Arc Cave have yet to be dated precisely, but even if they are from 33,000 years ago, we would still have to contend with the possibility that the paintings in the Cave of El Castillo in northern Spain are some 40,800 years old. To date, the identity of the artists is a matter of open debate. It is unclear whether these paintings provide early evidence of the presence of modern humans in Europe, or whether they were painted by Neanderthals. If indeed they were not painted by *Homo sapiens*, then another dividing line is blurred
In recent years, there has been a certain decline in support for the view that these caves represent the point of divergence. One of the reasons for this is the difficulty in dating the paintings, but the problem touches upon a far more fundamental issue: the inability to determine unequivocally that the ability to produce them is somehow unique to *Homo sapiens*
Image source: https://en.wikipedia.org/wiki/Lascaux#/media/File:Lascaux_painting.jpg

in *People of the Lake*: "Many animals modify their environment to some degree … but only humans impose their thoughts on the world around them in an entirely arbitrary and symbolic way" (1979, p. 185).

The cave paintings are important because they attest to abstract thought, which, in turn, attests to the cognitive ability to develop language. Once again, in the words of Richard Leakey: "There is no question that the evolution of spoken language as we know it was a defining point in human prehistory. Perhaps it was the *defining point* … Modern language made modern man" (1994, pp. 119–123).

Emphasizing the principle of evolutionary continuity, a number of scholars have attempted to draw a direct line from chimpanzees to humans. Language, however, poses a particular difficulty, even for these scholars, since it represents an immense evolutionary gap. Human language, especially syntax—which requires complex and exceptional cognitive ability—makes modern humans qualitatively and fundamentally different from every other creature. It is hard to determine when and how human language developed—whether it evolved from calls and vocalizations or from gestures.[7] Another problem concerns the development of a physical apparatus capable of producing the necessary sounds. Neither *Homo habilis* nor *Homo erectus* possessed such an apparatus, while the forms of communication they did use remain a mystery.[8] To continue this line of thought, early language was extremely basic, and it is hard to understand what evolutionary advantage it afforded in its earliest stages—certainly in relation to the amount of energy it required. Anyone who sits on a balcony from time to time and listens to the birds instinctively understands that they are constantly sending messages. In broader terms, there appear to be far more efficient ways of communicating. For example, chemical messages such as scents and pheromones are considered far more efficient than human language in terms of the quantity of information conveyed in relation to the amount of energy expended. For those who argue that language enables cooperation, one look at an ant colony will show that cooperation and joint problem-solving are possible even without language. Why then did language develop? It is easy to say that it offers evolutionary advantages, but it also has costs. Complex language requires a larger and more sophisticated brain, complicating birth

[7] For an extensive discussion, see (Hillert, 2021).

[8] Although *Homo habilis* and *Homo erectus* brains and larynges differed from our own, that does not mean that they lacked the ability to produce basic words, and it certainly does not suggest that they were unable to communicate in sign language.

and resulting in premature offspring, ill-equipped for survival in the world. The more we delve into this question the more perplexing it becomes—from claims that proto-language developed in parallel to our ability to make tools, some 2.5 million years ago, to the idea that language is the result of a genetic mutation that occurred around 100,000 years ago in Africa. According to American linguist Noam Chomsky, father of the generative approach to linguistics and one of the foremost intellectuals of the past 50 years, it was an abrupt and recent event that occurred "with no selectional or other pressures operating" (2016, p. 25). Chomsky, citing paleoanthropologist Ian Tattersall, suggests that "the abrupt and sudden event" probably lies "somewhere within the very narrow window of 50,000 to 100,000 years ago" (p. 3).[9] It is worth noting that in the first case (2.5 million years ago), language evolved slowly, in parallel to the evolution of *Homo*, and resulted in a significant increase in brain size and the development of *Homo erectus*—the first hominin to leave Africa, some 1.8 million years ago—while, in the second case (50,000 to 100,000 years ago), the result was the beginning of *Homo sapiens*' migration out of Africa. In both cases, there is a direct link between language, symbolic systems, and the ability of the human race, at various stages of its development, to expand beyond its boundaries and begin to dominate its surroundings.

We really don't know when or how language developed, just as we really don't know why we became bipedal. Some scholars have tried to identify findings that attest, indirectly, to the time of its appearance. More precisely, they attempt to determine when syntax—indicative of the existence of a linguistic structure—first developed. The working hypothesis of these researchers is that there is a connection between syntax and the structure and nature of tools fashioned by humans in various periods: a set of tools indicates the existence of syntax.[10] The logic behind this idea is that sophisticated tools are not always significantly better than stones, but the need for an orderly set of tools attests to the existence of an organized social structure that, in turn, attests to a cognitive ability to represent the world: "There are in fact remarkable commonalities between tool making and language, commonalities in the cognitive mechanisms that operate them and in their contribution to making *Homo sapiens* the only truly cultural animal" (Leakey & Lewin, 1979, p. 185).

[9] In this context, it is worth remembering the words of linguist Ray Jackendoff: "The problem of talking about the evolution of language in any detail is that there is no evidence. It's pure speculation" (Eakin, 2002).

[10] For an extensive discussion, see (Thibault et al., 2021; Gibson, 2012).

For a short time, the cave paintings allowed us to say, we are different; there is a clear dividing line between us and our rivals. It didn't last very long, however, since more recent findings indicate that *Homo neanderthalensis* created art and may even have performed burial rites. We are thus back where we started, asking ourselves what makes us, *Homo sapiens*, so unique. Leakey suggests that what sets us apart is our obsession with innovation: "Unlike previous eras, when stasis dominated, *innovation is now the essence of culture*, with change being measured in millennia rather than hundreds of millennia" (1994, pp. 93–94). This observation is, in itself, an admission that we don't really have a good answer to the question of what makes us unique and why we alone have survived. Were it possible to identify some gene or genetic system specific to *Homo sapiens* and were it possible to link that system to particular behaviors, such as the use of language, we could easily point to that as an evolutionary advantage (or at least a difference) with a genetic basis. The argument that it is our predisposition to innovation that sets us apart raises a number of questions: (1) Can a link be identified between this predisposition and a particular genetic structure? (2) Is the notion of a predisposition to innovation based on the findings themselves or on the way in which we think of humans in the hyper-technological modern era? (3) What type of findings could indicate a predisposition to innovation? At this point, it is worth noting: not only are most prehistoric tools lost, because they were made of perishable materials, but we don't really understand the uses of the tools we have found.

The assertion "innovation is the essence of culture" is completely detached from evolutionary discourse, so that the advantages this trait are supposed to have conferred lie in the realm of culture rather than genetics, which necessarily implies a particular cultural structure. If that is the case, how can we possibly say anything worthwhile about a culture that existed hundreds of thousands of years ago—and on the basis of fragmentary information at that? The answer is that there is nothing we can say for sure, which leaves a lot of room for interpretation—that is, relating to the past from the perspective of the present. There is no greater mistake a historian can make than examining the past through the lens of the present. If this is true of past decades and centuries, what is there to say about events that occurred tens and hundreds of thousands of years ago?

In an effort to identify the unique characteristics that set *Homo sapiens* apart from other species, many have attempted to take a step back in time, turning to hunter-gatherer tribes. Let's join them.

References

Braun, D. R., et al. (2019). Earliest known Oldowan artifacts at >2.58 Ma from Ledi-Geraru, Ethiopia, highlight early technological diversity. *Proceedings of the National Academy of Sciences, 116*(24), 11712–11717.

Bretas, R., Yamazaki, Y., & Iriki, A. (2020). Phase transitions of brain evolution that produced human language and beyond. *Neuroscience Research, 161*, 1–7. https://doi.org/10.1016/j.neures.2019.11.010

Chomsky, N. (2016). *What kind of creatures are we?* Columbia University Press.

Drummond-Clarke, R. C., Kivell, T. L., Sarringhaus, L., Stewart, F. A., Humle, T., & Piell, A. K. (2022). Wild chimpanzee behavior suggests that a savanna-mosaic habitat did not support the emergence of hominin terrestrial bipedalism. *Science Advances, 8*(50). https://doi.org/10.1126/sciadv.add.9752

Eakin, E. (2002, May 18). Before the word, perhaps the wink?; Some language experts think humans spoke first with gestures. *The New York Times.*

Gibson, K. R. (2012). Tool-dependent foraging strategies and the origin of languages. In K. R. Gibson, & M. Tallerman (Eds.), *The oxford handbook of language evolution* (pp. 339–349). Oxford Academic.

Harmand, S., et al. (2015). 3.3-million-year-old stone tools from Lomekwi 3, West Turkana, Kenya. *Nature, 521*(7552), 310–315.

Hillert, D. (2021). How did language evolve in the lineage of higher primates? *Lingua, 264*, 103158.

Leakey, R. E. (1994). *The origin of humankind.* London.

Leakey, R. E., & Lewin, R. (1979). *People of the Lake: Mankind and its beginnings.* Avon.

Milo, D. S. (2009). *The invention of tomorrow (Heb.).* Hakibbutz Hameuchad.

Ponce de León, M. S., et al. (2021). The primitive brain of early Homo. *Science, 372*(6538), 165–171.

Thibault, S., Py, R., Gervasi, A. M., Salemme, R., Koun, E., Lovden, M., Boulenger, V., Roy, A. C., & Brozzoli, C. (2021). Tool use and language share syntactic processes and neural patterns in the basal ganglia. *Science, 374*, 6569.

Wrangham, R. (2009). *Catching fire: How cooking made us human.* Profile Books.

CHAPTER 2

Gatherer-Hunters or Hunter-Gatherers: "The Killer Awoke Before Dawn"

(The Doors, The End)

The Hunting Hypothesis

A basic research question in anthropology is whether humans are predominantly gatherers or hunters.[1] This question is particularly relevant in light of the twentieth century, in which hunting, killing, and genocide became exceedingly efficient. We might also ask whether there is a connection between hominids becoming hunters and the emergence of *Homo sapiens*. For many scholars, the answer would appear to be in the affirmative. American anthropologist Keith Otterbein, for example, argued in *How War Began* (2004) that hunting and war go hand in hand. In order to understand this, we must look at the period preceding the appearance of *Homo*—2 to 4 million years ago. At the time, various regions in Africa were inhabited by *Australopithecus*—a bipedal ape that used tools and had a brain that still resembled that of a chimpanzee. The appearance of *Australopithecus* was linked to climatic change and the subsequent reduction of forested areas in favor of savannah. More prey than hunter, *Australopithecus* had to learn to cooperate in order to survive attacks by large animals. Otterbein suggests that it was this cooperation that later turned us into hunters. In other words, *Australopithecus*' hunting-related

[1] Since this question is far from settled, I have used the terms hunter-gatherer and gatherer-hunter interchangeably throughout the text.

Y. Ataria, *Genes, Technology, and Apocalypse*,
https://doi.org/10.1007/978-3-031-59197-6_2

cooperation was one of the central causes of the development of *Homo*—first *Homo habilis* and later *erectus* and *sapiens*.

Cooperation, in and of itself, did not turn us into hunters, but it was probably a precondition or is at least presented as inherent to that process in the scientific literature: those who fail to cooperate with the band will not survive. It is important to note the presence of two, conflicting, evolutionary pressures here. On the one hand, individuals, particularly males, must be loyal to the band that protects them. On the other hand, every male is in competition with other males for access to females. The question is, as always, which evolutionary pressure is stronger.[2]

Australopithecus, as a code name for a species that preceded *Homo habilis*, has a great deal to teach us about our own evolutionary roots. This species appears to have lived under the constant threat of predation by larger animals, perhaps also chased away from carrion, on occasion, by other scavengers. Being weaker than its predators, *Autsralopithecus* began to develop tools in order avoid face-to-face encounters with those predators—driving them off from a distance, for example, by throwing sticks and stones. This action required a change in the structure of the hand, to allow more complex use of tools, and perhaps a more sophisticated brain as well. Indeed, recent studies show changes in hominin hand structure and the use of more complex tools in the million years preceding the appearance of *Homo habilis*. Some have taken this idea a step further, arguing that the evolutionary pressures that led to tool use were rooted in the need to create weapons. These may have been defensive at first, but quickly (in evolutionary terms) became hunting implements. According to Otterbein, the hunting techniques of *habilis*—the origins of which can be observed in chimpanzees (walking in single file and lying in wait)—combined with the sophisticated technologies they employed, make this species the inventor of war. The more we read on the subject, the more appropriate it seems to put it differently, more radically: *habilis* did not invent war; war invented *habilis*, progenitor of the genus *Homo*. This idea is immensely significant, for if this is indeed so, then war is embedded in our genetic make-up—from the chimpanzee, through *Australopithecus*, to the tool-making, skill-honing *habilis* and the entire human family.

These tools, particularly bones and clubs, made us hunters, eaters of fresh meat, and enabled the development of *Homo*—first *habilis*, and then *erectus*. It was these clubs, it is often claimed, that allowed *Homo erectus* to

[2] For further debate see: (Clutton-Brock & Huchard, 2013).

leave Africa, reaching as far as modern-day China. The very same tools also served to eliminate those individuals who enjoyed the protection of the band but failed to cooperate.

When we speak of tool use, we are referring primarily to the hand axe, associated with meat-eating because it allows the hunter to cut skins and flesh and perform other tasks related to hunting and meat-eating. According to this approach, the use and development of tools is intrinsic to the transition of pre-*Homo* from prey to predator, hunted to hunter. Over the course of hundreds of thousands of years, *Homo erectus* developed more sophisticated hunting techniques, hunting larger animals and consuming greater quantities of meat—necessary for brain development. What emerges is essentially a feedback loop: hands that grip tools better, more sophisticated tools, larger brain and improved cognitive ability, cooperation skills, hunting, meat, and so on. The more *Homo* develops, the more the population grows, the more friction and competition for ever-larger hunting grounds increases, resulting in war.

In parallel to migration from Africa, *erectus* continued to evolve in the continent itself—evolution that entailed more sophisticated hunting techniques. As noted, the animals of Europe and Asia had not yet learned to fear *Homo erectus*, making hunting easier. The animals in Africa, on the other hand, were all too familiar with the fact that these creatures were predators as well as prey, and therefore kept their distance—leading to the development of even more sophisticated technologies, such as spears, bows, and other tools that allowed the hunter to strike from a distance; a more sophisticated cognitive system that made it possible to create such tools; and social systems that enabled greater cooperation. Viewed from this perspective, the development of *Homo erectus* was inexorably linked to hunting and meat-eating. This process led, some 400,000 years ago to the hunting of *big game*—the systematic pursuit of larger animals being a uniquely human behavior. At this stage, *Homo* entered a phase of rapid development, eliminating an increasing number of large animals, requiring significant territorial expansion. The result was increased tension and conflict between hunter-gatherer tribes, resulting in more wars and a genetic preference for warriors. Improved technology and hunting techniques, combined with climate-induced evolutionary pressure, made war more frequent.

Increasingly sophisticated cognition enabled the development of symbolic systems and group identification with a particular belief system, alongside opposition to other belief systems—effectively two sides of the same coin (and those who sense an attempt to find evolutionary justification for nationalism are not wrong). As we will see later in the chapter, the feedback loops between hunting and technological development are what enabled *Homo sapiens* to leave Africa for Europe and pretty quickly bring about the demise of the Neanderthals, but I am getting ahead of myself. For now, let us return to hunting.

In early prehistory, *Homo sapiens* appear to have lived in groups of 25 or so individuals, requiring an area of around 60 km^2 for their survival. At times, such groups were part of a larger tribe, numbering around 500 individual members. Even today, there is no consensus among scholars regarding the diet of ancient humans—a particularly important question in light of the fact that the development of a large brain would have required meat. As sedimentologist Gail Ashley put it, "We know that the increase in the size of the brain, just the evolution of humans, is probably tied to more protein" (Bates, 2016).[3] Some have argued that this does not necessarily mean that *Homo sapiens* hunted. It is possible, for example, that *sapiens* also scavenged—at least at certain stages of development.

Before going on, I would like to pay some attention to the difference between scavenging and eating fresh meat. Like us, baboons and chimpanzees are also omnivores, but will not touch carrion.[4] It is also important to note that chimpanzees generally eat gathered food separately but consume meat together, possibly indicating that meat-eating plays a role in the development of the individual as an integral part of a community. To continue this line of thought, to the extent that we were scavengers at various stages of our development and under certain conditions, the passage from scavenging to eating fresh meat is an important part of the human evolutionary story.

[3] The more we look into the hunting hypothesis, the more problematic it appears. It is unclear when and why we began to hunt large game. In terms of energy expenditure, gathering is a far more efficient and reliable way of acquiring protein, both in terms of the hunting process itself and in terms of the chances of success. Chimpanzees, for example, get all the protein they need in a half an hour of gathering termites. Moreover, large quantities of protein (more than 250 g of meat per day) can be toxic and lead to weight loss, because of the difficulty in digesting meat. Regarding the argument that protein is necessary for brain development, the key nutrient would appear to be fat rather than protein. Why hunt relatively lean animals then? Furthermore, anthropological studies of hunter-gatherer tribes have shown that most of the meat obtained by hunters is ultimately consumed not by them or their families, but by tribal leaders. Considered in this light the motivation behind hunting large game becomes increasingly unclear (Speth, 2010).

[4] For further debate see: (Stanford & Bunn, 2001).

In any event, it is important to understand that the development of *sapiens* was linked to meat-eating, but also when *Homo* (from *habilis*, through *erectus*, to *sapiens*) or pre-*Homo* (from the chimpanzee to *Australopithecus*) began to eat meat on a regular basis, whether fresh or carrion, whether the animals hunted were larger or smaller than the hunters, and what hunting technologies and techniques were used. The prevailing thesis, even today, is that although many animals eat meat, only humans (once again, perhaps this is precisely what made us human) regularly hunt animals larger than themselves, and that there is a connection (positive feedback loop) between meat-eating and brain development, between hunting large animals and cooperation, and between hunting large animals and the development of technology.[5] As noted, although decades have passed since the publication of Dart's work on the subject, many scholars continue to accept this premise. A classic example of this is Henry Bunn's "Meat Made Us Human" (2007). This approach appears to be deeply entrenched in science and culture.

Let us now focus on a hypothetical group of 25 individuals with a range of skills. One of the main characteristics of this group is its practice of sharing food, which imposes restraint. Ardrey argues that, "Above all we had to find means of inhibition in this brain of ours on behalf of risk-taking and cooperation, and deferred reward. The survival necessity of bringing the meat back home was a shattering development in the psyche of the primate male" (1976, p. 139). The individual gathers food and does not eat it immediately (unlike other apes), but adds it to the food obtained by another member of the group, to be eaten together later. This type of behavior seems to shed light on a number of important human traits. Deferring gratification forces hominids to plan and to extend their axis of time to the future and the past (recalling that it was worthwhile last time), which require complex abstract thinking and a well-developed imagination.

When an individual finds food, the temptation to eat it immediately and lie to the group is strong, since it is in the interest of individual members to make themselves stronger, in the context of internal struggles, including mating competition. This is a typical human trait, and numerous

[5] To cite one example of the ways in which this connection can be questioned, use of technology in the early stages of human development appears to have been entirely unrelated to male hunting: "Among chimpanzees, females and juveniles are the subsets that most frequently use tools, and most social transmission about tool use skills takes place between mother-offspring dyads" (Thompson et al., 2019, p. 9).

studies have shown both attempts to deceive and the way in which groups mercilessly punish those who try to hide information from the group. Every member of the group knows how they would like to behave if they could; hence, the need to keep an eye on other members and to develop the ability to spot “cheats”.

At the next stage, individuals begin to understand how they are viewed by others and “improve” their behavior to become better liars. In the process, the gaze of the other becomes a source of human anxiety. The prominent French philosopher Jean-Paul Sartre (1956), perhaps the person most identified with existentialism in the twentieth century, considered the “gaze” of the other a basic structure of human existence. The gaze of the other leads to the establishment of laws, as a means of protecting the group. Every infringement is met with an immediate reaction: “Modern gatherer-hunters are raised most rapidly and stormily when someone is discovered to have committed some kind of injustice, however small. The whole band may become involved in the outburst” (Leakey & Lewin, 1979, p. 137). Taken further, one might argue that culture has the power, at least to some extent, to temper primal human instincts and tendencies. Consider this simple assertion: we have an evil or at least problematic nature, but through culture, we may become sublime. Cultured humans are good because they have learned to control their basic nature. It is also possible, of course, that the opposite is true and we are good by nature but ruined by culture. Culture, according to one of the prevailing theories today, is what compensates individuals for the risks they take when they fight on behalf of the group.[6] In other words, although the group does defend its individual members, it is hard to explain self-sacrifice for the sake of the group from an evolutionary perspective that focuses on the individual. This is precisely where culture comes in—as the element that reinforces the compensation to warriors, first and foremost in relation to mating potential. This mechanism can explain, among other things, the glorification of soldiers in many societies. By making soldiers culture heroes, we make their sacrifice more worthwhile. Both theories—the reduction of everything to evolutionary mechanisms and the reduction of everything to cultural structures—are superficial and simplistic, but both serve the same purpose: to exonerate humankind.

[6] For an extensive discussion, see (Beckerman et al., 2009).

Cooperative hunting undoubtedly requires communication and tool use.[7] Gathering and conserving plants and seeds in carrier bags also require cooperation and tools. Unlike hunting tools, however, these are made of relatively perishable material and "There are, of course, no signs of this major technological revolution in the archaeological record" (Leakey & Lewin, 1979, p. 127). The hunting bias—which stems not only from the surviving physical evidence but also from a male perspective and the experience of the twentieth century, in which murders on a vast scale became almost a matter of routine—tells a limited story.[8] Indeed, some argue that the carrier bags, associated primarily with females, are in fact the technology through which human nature should be considered. Obviously, a lifestyle that included gathering and the use of tools for carrying and preservation would have made a significant contribution to the development of cognitive ability. We spoke earlier of the importance of protein to brain development, but termites—an abundant resource, gathered by females—also provide protein. The problem is that if termites were the source of our larger brains, rather than fresh meat, little if anything remains of the hunting hypothesis.

[7] "Conventional wisdom in human evolutionary studies had long assumed that the origins of hominin sharp-edged stone tool production were linked to the emergence of the genus *Homo* in response to climate change and the spread of savannah grasslands" (Lewis & Harmand, 2016).

[8] Recent studies have shown that women probably also took part in hunting (Hass, 2020). The idea of hunting as a determining factor in the expansion of the neocortex and higher intelligence emerged from a conference held in Chicago in 1966, at which scholars gathered to discuss modern "foraging peoples" (hunter-gatherers). The volume that followed included the chapter "The Evolution of Hunting", by Sherwood Washburn and Chet Lancaster, in which the authors suggested that hunting is "among the most fundamental of human behavioural adaptations" (Stanford & Bunn, 2001, p. 4), noting the influence of communication and coordination inherent to big game hunting on the development of intelligence and the expansion of the neocortex. In placing such emphasis on hunting—a primarily male activity in modern societies—they ignored the role of females in the evolution of the human brain. Some took issue with Washburn and Lancaster's hypothesis elicited assumption that meat constituted a significant part of the diet of early humans, while others argued that their conclusions were rooted in "male-based science". Ironically, most of the scholars attending the conference agreed that meat played a relatively small role in the diets of modern foragers. For further debate see: (McGrew, 2014). In any case I agree that "Any account of hominid evolution that concentrates only on males, or only on females, or suggests that any specific age-sex class is responsible for hominid evolution, is defectiv. Both 'Man the Hunter' and 'Woman the Gatherer' analyses frequently suffer from this problem." (Tooby & DeVore, 1987, p. 192).

Thus, when the focus is shifted to gathering and preservation, a different narrative emerges. At the same time, however, we cannot ignore the skills that make humans such good hunters. For example, only humans can throw a basketball in a hoop, pass a football on a pitch, or hit a tennis ball at a speed of 200 km/h. Our closest genetic relatives have no such skills; this brings us back to our use of tools and the link between technology, reason, cooperation, and hunting, as the basis of the killer ape theory.

Since cooperation between humans is an incontrovertible fact, there must be some underlying trait that explains how, in the words of American anthropologist S. I. Washburn, "The very same actions that caused man to be feared by other animals led 'to food-sharing, more cooperation, and economic interdependence'" (Ardrey, 1976, p. 13). As we have seen, scholarship on this subject is less fact-based research than evolutionary narrative through the prism of the political, cultural, and social reality of the second half of the twentieth century reality. Here too Ardrey is the perfect example: "In all our skills, as in all our philosophies, something seems lacking, so much more skillful are we at implementing our antagonisms than our affections" (p. 9), and furthermore, "*If among all the members of our primate family the human being is unique … it is because we alone through untold millions of years were continuously dependent on killing to survive*" (p. 11 [Emphasis in the original]). In the transition from jungle to savannah,[9] when humans could no longer flee to the trees, they passed a point of no return:

> An evolutionary moment had to come for the evolving hominid when his life became more dependent on the successful kill than the secure escape … The renunciation of the trees meant a reversal of primate psychology: we could flee only so far and so fast; then we had to fight … We passed a point of no return in our commitment to the ancient, hazardous savannah. (Ardrey, 1976, p. 64–65)

According to Ardrey, the hunting hypotheses also explains human gender structure, the female orgasm, concealed ovulation, anatomical differences between males and females, copulation throughout the ovarian

[9] Recent findings have shown that such a transition never actually took place: "However, it has become apparent that before 2 to 3 million years ago (Ma), early hominins occupied woodland environments, not open or even bush-savannah environments … and that although hominin hind limbs rapidly became adapted for terrestrial bipedalism, they retained long grasping forelimbs, which are more obviously relevant in an arboreal context" (Thorpe et al. 2007).

cycle, and variable mating positions. All of these are related to the need of the female to bind a male to her and ensure that she and her offspring receive their portion of meat, without arousing other males, in order to preserve group structure and cooperation. Such relationships would be contingent on the ability of the male to hunt and then wait to share food.

Ardrey further argues that the ability of to hold a weapon was the primary force behind erect bipedalism:

> To carry a weapon demanded free hands. To free our hands, a degree of erect posture and bipedal movement, however awkward, became a necessity. And so every selective pressure favored those of us who most reliably could stand erect and move on two feet, *just as it favored those newly freed hands that could most ably wield a weapon.* (p. 43)

In this, Ardrey appears to have remained faithful, at least in part, to Darwin, who argued that "the hands and arms could hardly have become perfect enough to have manufactured weapons, or to have hurled stones and spears with true aim, as long as they were habitually used for supporting the whole weight of the body" (Darwin, 1871, pp. 140–142). Although ostensibly similar, the difference between Ardrey and Darwin on this point is significant. While Darwin posits that erect posture *preceded* the use of weapons, Ardrey cites the use of weapons as the *cause* of erect posture. In general, Ardrey tends to seek teleological explanations. His assertion that "aside from the weapon itself, there is no element in human culture that has persisted so changelessly for so long" (Ardrey, 1976, p. 164) thus comes as no surprise.

Even if we reject this hypothesis, we cannot deny that it attempts to explain a variety of phenomena. At the same time, we must not ignore the fact that it is a male-centered hypothesis devoid of any real scientific basis, the true purpose of which, I would argue, is to explain the killing fields of the twentieth century (from which it, in fact, derives). We don't know why we became bipedal, what led to more sophisticated tool use, or why language developed. These questions are crucial to the hunting hypothesis, yet we are far from understanding any of them let alone all of them together. Indeed, it is readily apparent, despite Ardrey's assertion that "things do not evolve with purpose" (p. 40), that there is a clear purpose, manifested in Auschwitz and the atomic bomb. In other words, books like Ardrey's, Pinker's (on which, more below), and Harari's teach us a great deal about white, Western, male self-perception in the second half of the twentieth century, but little or nothing about the past.

The Origin of the Hunting Hypothesis

In 1924, a number of extraordinary fossils were discovered in a cave in South Africa. One of them, the skull in Fig. 2.1, intrigued a student by the name of Josephine Salmons, who brought it to the attention of Australian paleoanthropologist Raymond Dart, then at the beginning of his career.

Dart examined the skull and noticed that it was larger than that of a baboon or chimpanzee, and that the opening (foramen magnum) through which the spinal cord connects to the brain was lower, indicating upright locomotion. Dart described the finding, which came to be known as Taung Child skull, as a new species—a species he named *Australopithecus Africanus.* According to Dart, *Australopithecus* lived in southern Africa some 2.5 million years ago, and constitutes the missing link between chimpanzees and modern humans (Dart, 1925). The scientific community rejected this claim until the 1950s. Eurocentrism had led scientists to prefer another finding—a skull discovered in England in 1912 (Piltdown Man). The desire to believe that humans had originated in Europe and that the development of the brain had preceded bipedalism, prevented them from recognizing that it was a fake—consisting of a modern human cranium combined with an altered orangutan mandible and filed-down teeth. It was thus only after the main finding supporting the theory that

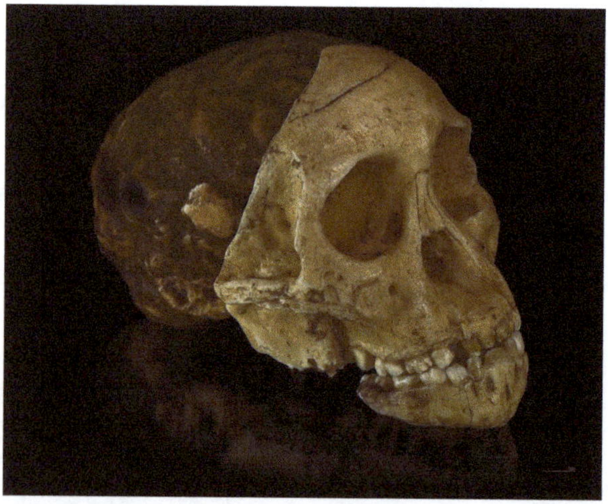

Fig. 2.1 Taung Child—*Australopithecus africanus*

humans had originated in Europe had definitively been proven a fake that Dart's position that the *Australopithecus africanus* skull attests to humankind's African origins (vindicating Darwin) was widely accepted.

Dart wondered how this species survived without ready access to food. Examining the skull, he discovered a small hole on the right-hand side and hypothesized that it had been made by an adult *Australopithecus*, in order to eat the child's brain. The cave in which the skull was found also contained fossils of turtles, birds, and baboons, which Dart interpreted as food remains left behind by *Australopithecus*. He thus suggested that it was by hunting small animals that *Australopithecus* managed to survive in a hostile and unfavorable environment. Furthermore, he argued, it was hunting that allowed this hominin to embark on the evolutionary path ultimately leading to modern humans.

In 1953, Dart presented an even more compelling theory, based on findings discovered over the years—that *Australopithecus* hunted large game and was, first and foremost, a hunter (Dart, 1953), driven out of the jungles and into the savannah in search of fresh blood. One of the problems with Dart's theory is that no tools have been found from the period in question, by means of which *Australopithecus* is supposed to have hunted large game. Dart postulates that the tools used by *Australopithecus* were made of bone.

Dart's theory, rejected in the 1920s, was warmly embraced in the apocalyptic climate of the second post-war period. Attempts to understand how we arrived at a situation in which hundreds of thousands of deaths in a single night (e.g., the night of the firebombing of Tokyo on the night of 9–10 March 19, 1945) are considered a reasonable price to pay appear to have encouraged scientists to accept explanations that describe humankind in the most violent possible terms. Theories of this kind are warmly received in popular culture, since the general public also seeks some way of explaining the horrific number of people murdered in the twentieth century—how ordinary people could be capable of such cruelty. Indeed, in the aftermath of the Second World War, the apocalypse did not seem so far off. Dart, who had portrayed humankind as a particularly violent species from the very beginning, was suddenly perceived as having expressed a profound understanding of human nature. The evidence for this theory, however—and this is the central argument of this section—was not discovered in Africa, but in Auschwitz and Hiroshima. In the 1950s, some scholars further refined Dart's ideas, suggesting that large game hunting requires group cooperation, from which they concluded that hunting lies at the root of human cooperation. Scholars also developed the theory that males went on hunting expeditions while females were left behind in a protected location. Males stood upright in order to be able to grasp tools in their hands, while females did so in order to care for their offspring. It is easy to see that these scholars—almost all men—based their theories on their own reality rather than on any prehistoric findings.

Over the years, Dart's theory gained currency and was taken to an even greater extreme both in the scientific community and in popular culture: humans are born killers and cannibals. In the 1950s, Dart presented a comprehensive theory whereby *Australopithecus* hunted not only competing species, but also members of its own species, which, Dart argued, is what makes us unique, sets us apart, turns us into a dominant species.[10]

At this stage, it is worth asking on what evidence did Dart base his theory? Dart examined damage to the Taung Child skull and concluded that it attested to an attempt to open the skull in order to eat the brain. There are, however, other possible explanations for the type of damage found in this and other skulls: in one case, the damage corresponded exactly to the fangs of a leopard, in another to an eagle's beak, and in yet another appears to have been caused by a rock pressed against the fossilized skull.

Moreover, no evidence of tools—including those made of bone—has been found. *Australopithecus* would thus have failed to survive even as a scavenger. Indeed, it is hard to imagine a group of *Australopithecus* chasing away a pride of lions with their bare hands. There are also other problems with Dart's theory. For example, the fact that chimpanzees also eat meat means that meat-eating is neither unique to humans nor could it have made us what we are. What is more, when we examine modern hunter-gatherers, there is no evidence that they rely on large game or cannibalism.

[10] On the state of research at the time, Leroi-Gourhan (1993, p. 17) wrote: "To sum up, one might say that in 1950 the image of the primitive human was undergoing a far-reaching transformation. As in all times of change the various positions were not always very clear-cut, and the finest scholars were sometimes torn between contradictory hypotheses."

Lady chimps will mate for meat. Source: ©CristinaGomes (I would like to express my gratitude to Cristina Gomes for granting permission to use this picture)

American anthropologist Matt Cartmill (1993) argues that Dart's theory was wholly unsupported by physical evidence and stemmed entirely from his own belief system: Dart wanted to believe that humans are born killers and found the evidence he was looking for. Cartmill suggests that Dart's approach echoes the Christian story of Adam's Fall,[11] retold in scientific terms, combined with Odyssean warrior mythology: the reduction of arboreal areas led chimpanzees to descend from the trees, leave the forests, and learn to survive in the savannah (the Fall from Eden), to which end they had to develop as supreme hunters (the warrior myth). American anthropologist Robert Sussman (1999) suggests that *Australopithecus* was hunted rather than hunter. The findings cited by Dart are consistent with this theory.

It seems that after the Second World War—in which it was proven, once again, what humans are capable of doing to one another—there was greater receptiveness to Dart's interpretation. His theory affords a degree of comfort and even exoneration, because the horrors of the Second World War represent what Dart defined as the "state of nature". It is not that Dart's theory was itself a product of the Second World War—it in fact

[11] At this point, we may indeed suggest that Western cultural discourse is shaped by structural trauma (see Ataria, 2017).

echoes the Hobbesian view of the "state of nature" of humankind as a "war of all against all"—but that it certainly suited the prevailing mood in the scientific community and the general public at that time and was thus easily accepted: humans are natural-born killers. In theory, the Second World War should have weakened belief in the Hobbesian idea of the state as a bulwark against the "nasty, brutish, and short" state of nature, since it was fought between sovereign powers. The opposite occurred, however, and the war was widely perceived as a reversion to the state of nature—a primitive, barbaric state to which Hitler had reduced Germany, resulting in a war of all against all. Consequently more accords and cooperation between powers are required in order prevent the state of nature from re-emerging. According to this approach, Auschwitz is the result of a certain state of nature.

Noble Savage or Born Killer?

For some 2 million years, humans lived in tribes of hunter-gatherers. If we accept the idea that our behavior is influenced by our genes, then the conditions in which *Homo* evolved, ultimately becoming modern humankind, is of great importance.

Before presenting the data itself, it is worth noting that there is no concrete evidence behind these theories, which are, in fact, entirely a matter of conjecture:

1. Knowledge of prehistoric societies (before written records) is by no means certain. French anthropologist Claude Lévi-Strauss (1952, 1971), one of the founders of structuralism, argued that changes imperceptible to Western eyes may have been revolutionary from the perspective of the societies that experienced them. Relating to present-day tribes as faithful representatives of communities that lived tens and hundreds of thousands of years ago is problematic, to say the least.
2. Prior to the agricultural revolution, there were countless tribes of gatherer-hunters that differed greatly from one another, thus making any attempt to describe tribes of "hunter-gatherers" in the same way simplistic and wrong.
3. Regarding the study of modern-day gatherer-hunters, there is no doubt that most have come into contact with agricultural populations for thousands of years. It is thus unclear what observing them

in the present actually teaches us about circumstances before the agricultural revolution—a problem undoubtedly exacerbated in recent centuries.[12]

4. Most hunter-gatherer tribes today live in relatively harsh climates, which naturally affect both individual and group behavior.

Canadian psychologist Steven Pinker, one of the most influential public intellectuals in the world in recent decades, is a popular representative of the Hobbesian view that humankind's natural state is one of war of all against all, and it is the political state that ensures peace. Pinker's views appear to be supported by numerous and rigorous studies. A broad survey, including tribes from a wide variety of cultures and geographical regions, reveals that some 15% of the population and a quarter of the males (between 7 and 50%) in gatherer-hunter societies die in violent conflicts, and fighting is a common phenomenon (Goldstein, 2011).[13]

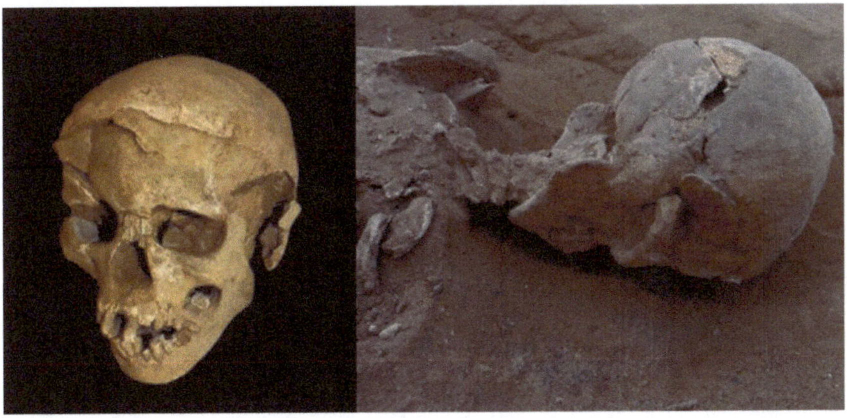

Photograph of the skull of an individual who died a violent death, from the site of Nataruk, Turkana, Kenya (photos by Marta Mirazón Lahr)

[12] See (Glowacki et al., 2020).

[13] For an extensive discussion, see (Diamond, 2012). In any event, the problem is not only that the difference between tribes in various periods is immense, but also that the various methodological approaches lead to completely different conclusions regarding the same tribes at the same points in time (Allen & Jones, 2014). Moreover, since interpretation of the physical evidence is far from certain and findings from earlier periods are scant, an assertion regarding a given fossil may assume disproportionate significance.

American anthropologist Brian Ferguson (2013) argues, however, that all of the data that Pinker relies on can be interpreted in other ways. And indeed, methodologically speaking, it is often difficult to determine the cause of death. Altogether, as we have seen, it is hard to say virtually anything about hunter-gatherer tribes prior to the agricultural revolution.

Although Ferguson does not completely refute all of Pinker's claims, he presents them in a new light—as do other studies. To put things a little more bluntly, from a scientific point of view, Pinker's work is embarrassing. He claims to take an evolutionary approach, yet none of the sites he focuses on is older than 14,000 years. Pinker knows full well that in evolutionary terms, humans were shaped far earlier than that, rendering the findings he cites completely irrelevant to the frame of reference he himself adopts—that of evolutionary psychology. Continuing this line of thought, Graeber and Wengrow (2021) argue that Pinker simply ignores a significant part of the evidence. Like Pinker, Ferguson too tends to ignore findings that indicate that hunter-gatherers did not live in peace and harmony until the agricultural revolution.

American anthropologist Douglas Fry (2013) suggests that Pinker's Hobbesian point of departure—that human history prior to the agricultural revolution was rife with war and death—is unfounded. At the same time, however, it can hardly be claimed that hunter-gatherers lived together in peace and it was culture (civilization, progress) that upset the harmony between humans and nature and made them into killers. It is hard to ignore the fact that Fry's conclusions are the product of methodological choices that some might even call Rousseauian (the "Peacenik" approach).

Israeli military historian Azar Gat (2006) focuses on Aboriginal hunter-gatherer tribes in Australia—a region in which agriculture did not develop prior to European settlement in 1788. Based on anthropological research, he argues that there was considerable inter-tribal violence, and that many males were killed as a result. Gat is of the opinion that Australia has a great deal to teach us about conditions among pre-agricultural gatherer-hunters, although it is hard to say how much of this is based on actual scientific research and how much is simply an attempt to find support for preconceived notions.[14] In light of the reservations presented above, it is worth noting that although the information scholars are able to provide regarding the lives of gatherer-hunters over the course of two million years is

[14] Indeed, despite Gat's efforts to justify this focus, we cannot ignore the fact that the tribes in that area are known to be particularly aggressive, regardless of factors such as population density, raising serious questions about his choice of region.

somewhat limited, the presence of violence is a certainty. Any attempt to portray the lives of hunter-gatherers as peaceful and war-free must ignore archaeological findings. Indeed, most scholars tend to think of the development of gatherer-hunter tribes in quite violent terms—always focusing on the male, while ignoring the role of women in human evolution (Fisher, 1979; Le Guin & Haraway, 2019).

In any event, just as there is no reason to embrace the Hobbesian view regarding humankind's natural state (humans as hunters), there is no reason to embrace that of Rousseau either—that humans are noble savages (humans as gatherers) spoiled by civilization, fallen from grace into a technological state of being (*technics*). Put another way, just as the concept of competition does not necessarily imply war, the concept of cooperation does not necessarily imply peace. The respective theories are rather simplistic. When the stakes are high and the chances of victory are relatively good, we are capable of fighting, while, in other circumstances, we are capable of pursuing peace. We are not genetically compelled to do either of these things. Indeed, there is something ridiculous about studies that attempt to associate a human propensity for violence with particular genes. Not surprisingly, such studies tend to present the prejudices of Western scholars regarding tribes they simply do not understand.

Of course, if our point of departure is Hobbesian, we will find objective evidence (so-called) that indeed, we live in a Hobbesian world and must therefore adopt appropriate policies (such as the nuclear strategy during the Cold War). This, in turn, makes the world increasingly Hobbesian and leads scholars to conclude that it was always so—that we are, above all, hunters—further contributing to the vicious circle.

The English philosopher Thomas Hobbes (1588–1679) influenced the Scottish historian and philosopher David Hume (1711–1776), who influenced the economist (and philosopher) Adam Smith (1723–1790), who influenced the cleric (and political economist) Robert Malthus (1766–1843), who influenced the father of the theory of evolution, Charles Darwin. Thus, the theory of evolution—perhaps the most important scientific theory and certainly one that shapes our entire way of thinking today—was influenced by the scientifically unproven Hobbesian approach (Ridley, 1996). One might conclude that Darwin's scientific theory is also driven by the idea of a natural state of war of all against all, but Darwin often refers to cooperation within species.

The nineteenth-century British sociologist and philosopher Herbert Spencer, one of the founders of social Darwinism, believed that it is better not to help the weak, because aggressive competition is the key to the development of our species. Spencer was of this opinion long before Darwin published his theory of evolution. In the theory of evolution, Spencer found scientific confirmation of his political convictions, although that appears to be precisely what he was looking for, having arrived at his conclusions on the basis of his own preconceived notions.

As anthropologist David Graeber and archaeologist David Wengrow have shown in *The Dawn of Everything* (2021), even after all these years, the points of departure (Hobbesian or Rousseauian – see Fig. 2.2) for the interpretation of every finding remain fundamentally unchanged. The

Fig. 2.2 Hobbes and Rousseau—simplistic views of human nature
The study of human nature is driven more by myth than by hard evidence. For example, we take for granted the idea that war is more ancient than humankind is, or that the state of nature is war of all against all, but is that indeed the case? Is there compelling evidence that humans have always been in a state of war, whether inter-tribal (e.g., over territory or resources) or intra-tribal (e.g., over females)? Is the way of life developed by the genus Homo over the course of millions of years, that of the gatherer-hunter, fundamentally Hobbesian? Rousseau's idea of the noble savage is equally unfounded. Gatherer-hunters did not live in perfect harmony with nature (the humans who first invaded America soon destroyed most of the continent's large mammals), only to be ruined by Western civilization

situation has actually worsened over time. Whether we embrace Rousseau or Hobbes, the end point remains the same. The result is a sense of the end of history—as if things could not have turned out otherwise.

Despite their critical approach, Graeber and Wengrow also *operate within the framework of the same metaphysics that views the beginning as determining the end*. What they demonstrate in their remarkable study is that there are other possible beginnings, but, like Nietzsche, they fail to question the very idea of a fundamental connection between the evolutionary conditions in which we developed and what we are today. Like others, Graeber and Wengrow look to the past in order to outline a new future and, in this sense, remain trapped within the same paradigm:

> What if we treat people, from the beginning, as imaginative, intelligent, playful creatures who deserve to be understood as such? What if, instead of telling a story about how our species fell from some idyllic state of equality, we ask how we came to be trapped in such tight conceptual shackles that we can no longer even imagine the possibility of reinventing ourselves? (p. 9)

Here we find the profound problem not only with Graeber and Wengrow, but with everyone, no matter how critical, trapped within the Western metaphysics: the desire for ends and beginnings, the idea that there is a fundamental, if not deterministic connection between them: "If 'what is the origin of inequality?' is not the biggest question we should be asking about history, what then should it be?" (p. 25).

Let us return for a moment to Dart, and to Ardrey who introduced Dart's views into public discourse. We now know that *Australopithecus* was a vegetarian (perhaps consuming meat occasionally, but definitely not hunting) and most of Dart's claims are entirely without basis: there is no reason to believe that it was the "thirst for blood" that brought our ancestors to the savannah. Does this then put to rest the idea that we are born killers? On the contrary, many scholars today argue with even greater conviction that there is nothing that distinguishes us from the chimpanzee, which is also a killing machine. The theoretical speculation presented as a necessary "scientific" explanation, whereby humans and chimpanzees exist along the same evil and morally flawed continuum, not only casts doubt on the notion of free will, but renders it superfluous.

References

Allen, M. W., & Jones, T. L. (2014). *Violence and walfare among hunter-gatherers.* Routledge.

Ataria, Y. (2017). *The structural trauma of western culture: Toward the end of humanity.* Palgrave Macmillan.

Ardrey, R. (1976). *The hunting hypothesis: A personal conclusion concerning the evolutionary nature of man.* Atheneum.

Bates, T. B. (2016, March 9). Early human habitat, recreated for first time, shows life was no picnic. *Rutgers Today.* https://www.rutgers.edu/news/early-human-habitat-recreated-first-time-shows-life-was-no-picnic

Beckerman, S. Erickson, P. I., Yost, J., Regalado, J., Jaramillo, L., Sparks, C., Iromenga, M., & Long, K. (2009). Life histories, blood revenge and reproductive success among the Waorani of Ecuador. *Proceedings of the National Academy of Sciences of the United States of America, 106*(20), 8134–8139.

Bunn, H. T. (2007). Meat made us human. In P. S. Ungar (Ed.), *Evolution of the human diet: The known, the unknown, and the unknowable* (pp. 191–211). Oxford University Press.

Cartmill, M. (1993). *A view to a death in the morning: Hunting and nature through history.* Harvard University Press.

Clutton-Brock, T. H., & Huchard, E. (2013). Social competition and selection in males and females. *Philosophical Transactions of the Royal Society B: Biological Sciences, 368*(1631), 20130074.

Dart, R. (1925). Australopithecus africanus the man-ape of South Africa. *Nature, 115*, 195–199.

Dart, R. (1953). The predatory transition from ape to man. *International Anthropological and Linguistic Review, 1*(4), 201–217.

Darwin, C. (1871). *The descent of man, and selection in relation to sex.* London.

Diamond, J. (2012). *The world until yesterday: What can we learn from traditional societies?* Viking.

Ferguson, B. (2013). Pinker's list. In D. P. Fry (Ed.), *War, peace, and human nature: The convergence of evolutionary and cultural views* (pp. 112–131). Oxford University Press.

Fisher, E. (1979). *The carrier bag theory of evolution. Womans creation – sexual evolution and the shaping of society.* McGrew Hill Book Co.

Fry, D. P. (2013). War, peace, and human nature. In D. P. Fry (Ed.), *War, peace, and human nature* (pp. 1–21). Oxford University Press.

Gat, A. (2006). *War in human civilization.* Oxford University Press.

Glowacki, L., Wilson, M. L., & Wrangham, R. W. (2020). The evolutionary anthropology of war. *Journal of Economic Behavior & Organization, 178*, 963–982.

Goldstein, J. S. (2011). *Winning the war on war: The decline of armed conflict worldwide*. Penguin.

Graeber, D., & Wengrow, D. (2021). *The Dawn of everything: A new history of humanity*. Penguin.

Hass, R. (2020). Female hunters of the early Americas. *Scientific Advances, 6*(45), eabd0310. https://doi.org/10.1126/sciadv.abd0310

Le Guin, U. K., & Haraway, D. J. (2019). *The carrier bag theory of fiction*. Ignota Books.

Leakey, R. E., & Lewin, R. (1979). *People of the Lake: Mankind and its beginnings*. Avon.

Leroi-Gourhan, A. (1993). Gesture and speech (A. B. Berger, Trans.). MIT Press.

Lévi-Strauss, C. (1952). *Race and history*. UNESCO.

Lévi-Strauss, C. (1971). Race and culture. *International Social Science Journal, XXIII*(4), 608–625.

Lewis, J. E., & Harmand, S. (2016). An earlier origin for stone tool making: Implications for cognitive evolution and the transition to homo. *Philosophical Transactions of the Royal Society B: Biological Sciences, 371*, 20150233. https://doi.org/10.1098/rstb.2015.0233

McGrew, W. C. (2014). The 'other faunivory' revisited: Insectivory in human and non-human primates and the evolution of human diet. *Journal of Human Evolution, 71*, 4–11.

Otterbein, K. F. (2004). *How war began*. Texas A&M University Press, College Station.

Ridley, M. (1996). *The origins of virtue*. Viking.

Sartre, J.-P. (1956). *Being and nothingness: An essay on phenomenological ontology*. Philosophical Library.

Speth, J. D. (2010). *The paleoanthropology and archaeology of big-game hunting: Protein, fat, or politics?* Springer.

Stanford, C. B., & Bunn, H. T. (Eds.). (2001). *Meat-eating and human evolution*. Oxford University Press.

Sussman, R. W. (1999). The myth of man the hunter, man the killer and the evolution of human morality. *Zygon, 34*(3), 453–471.

Thompson, J. C., Carvalho, S., Marean, C. W., & Alemseged, Z. (2019). Origins of the human predatory pattern: The transition to large-animal exploitation by early hominins. *Current Anthropology, 60*(1), 1–23.

Thorpe, S. K., Holder, R. L., & Crompton, R. H. (2007, June 1). Origin of human bipedalism as an adaptation for locomotion on flexible branches. *Science, 316*(5829), 1328–1331. https://doi.org/10.1126/science.1140799

Tooby, J., & DeVore, I. (1987). The reconstruction of hominid behavioral evolution through strategic modeling. In Warren G. Kinzey (Ed.), *The evolution of human behavior: Primate models*. SUNY Press.

CHAPTER 3

Back to the Chimpanzee

The Common Chimpanzee: Neighborhood Bully or Spiritual Forebear of Modern Humans?

Many scholars believe that the best way to understand modern humans is to observe chimpanzees. Jared Diamond is the author of the Pulitzer Prize-winning *Guns, Germs, and Steel* (1997) and one of the most prominent voices in popular discourse of the past 30 years. His work draws on the fields of anthropology, evolution, biology, and history. Diamond's point of departure is that humans and chimpanzees belong to a single family, because we "share over 98 percent of our DNA with chimps" (2006, p. 169) (the similarity between humans is 99.5%), or, more precisely "We differ in 1.6 percent of our DNA from [the common and pygmy chimps] … and diverged from their common ancestor around seven million years ago" (ibid., p. 21). To put it another way, "the chimpanzees' closest relative is not the gorilla but humans" (ibid., p. 23).

Diamond concedes that we have no way of knowing what caused the leap that made humans human and set them apart from other apes. In the absence of a better explanation, he accepts the theory that the difference lies in language. Although "chimpanzees, gorillas, and even monkeys are capable of symbolic communication not dependent on spoken words" (ibid., p. 54), Diamond writes,

Y. Ataria, *Genes, Technology, and Apocalypse*,
https://doi.org/10.1007/978-3-031-59197-6_3

> language lets us communicate with each other far more precisely … lay joint plans, teach one another … learn from what others experienced elsewhere or in the past … store precise representations of the world within our minds, and encode and process information far more efficiently than can any animals. (ibid., p. 141)

The genetic relationship between humans and chimpanzees explains our desire to understand them—a desire that is heightened by the fact that all of the intermediate species between chimpanzees and modern humans are extinct. Chimpanzees, we believe, will reveal some hidden truth about ourselves, particularly with regard to our propensity for violence. As Michael Ghiglieri, a biologist and protege of Jane Goodall, put it, chimpanzees allow us to claim that human violence "instead of being an unfortunate by-product of civilization … has much deeper roots" (Ghiglieri, 1999, p. xiii)—particularly with regard to the human male. It is interesting to note that neither Ghiglieri nor other scholars explore the possibility that although male chimpanzees may be violent today, that does not necessarily mean that they were violent 6 million years ago. What evidence is there that chimpanzees have always been violent? Just as *Homo* developed and changed over time, chimpanzees may have done the same (Sussman, 2013), Furthermore, most researchers of chimpanzees ignore the fact that chimpanzee habitats have been significantly reduced in recent decades, so that the troops we observe are in a state of distress—to which violence is a possible response (Ferguson, 2023).

Ghiglieri grounds his claims in sex and reproduction, especially the genetic results of competition between males for opportunities to mate, arguing that "the single most useless—and dangerous—approach one could take in trying to explain human violence is to look only at nurture, while ignoring how sexual selection has sculpted the evolutionary software programming the differences between men and women" (Ghiglieri, 1999, p. 12). Chimpanzees offer an example of this: "Wild chimps reveal the natural contexts of territoriality (Fig. 3.1), war, male cooperation, solidarity and sharing, nepotism, sexism, xenophobia, infanticide, murder, cannibalism, polygyny, and mating competition between kin groups of males—behaviors that have evolved through sexual selection" (p. 176).[1]

[1] It is important to note that this is a simplistic theory even from an evolutionary perspective, since greater competition does not necessarily lead to greater violence. Competition can manifest itself in many different ways, and need not entail direct violence. I am grateful to Ayelet Shavit for this observation.

Fig. 3.1 Chimps on Patrol
Studies conducted in the late twentieth century found that chimpanzees that kill other chimpanzees do not necessarily do so for reasons of social hierarchy. Most such cases of violence were found to take place within the male group and were generally directed against a weaker male (a "scapegoat"). Such murders often involve genital and facial mutilation. In this context, it is interesting to recall Darwin's observation that "the mental powers of the higher animals do not differ in kind, though greatly in degree, from the corresponding powers of man" (Darwin, 1998, p. 560) (I would like to express my gratitude to Mitani for granting permission to use the photograph)

Influenced by Chomsky and his theory of universal generative grammar, Ghiglieri finds the roots of cultural structure in our natural instincts: "Our cultural symbols drive us this hard because our instincts command us to conform and to identify with our group" (Ghiglieri, 1999, p. 64).[2] He argues that culture ultimately rests on natural foundations and explains both group identification and the desire to destroy other groups, which are two sides of the same coin: a cultural structure rooted in instinct. Studies of hunter-gatherer tribes have shown that, like chimpanzees, males remain with their group of origin, while females move from group to group. Ghiglieri argues that this phenomenon of retaining males "set the stage for cooperative ape warriors" (ibid., p. 71).

[2]This might be a good time to see how Ghiglieri describes himself on his own website: "Michael P. Ghiglieri grew up at Lake Tahoe, Nevada as the great-grandson of a Forty-niner, served as a US Army platoon sergeant during the Vietnam era, then earned his Ph.D. in Ecology in 1979."

In their attempt to understand how we came to shed so much blood—a question greatly heightened by the events of the twentieth century—Diamond and Ghiglieri depart from an evolutionary perspective, or, more precisely, a deterministic, genetic evolutionary perspective. Their approaches differ slightly, but the end result is the same: Auschwitz. According to Diamond, genocide is hardly a recent phenomenon. Hunter-gatherers, early agricultural societies, and every human community that has ever existed have hunted, murdered, and exterminated: "Evidently, genocide has been part of our human and prehuman heritage for millions of years" (2006, p. 297). Furthermore, Diamond writes:

> Chimps also share xenophobia with us: they clearly recognize members of other bands as different from members of their own band, and treat them very differently … Common chimps already carried out planned killings, extermination of neighboring bands, wars of territorial conquest, and abduction of young nubile females. If chimps were given spears and some instruction in their use, their killings would undoubtedly begin to approach ours in efficiency. (ibid., p. 294)

Despite the differences between their approaches, Ghiglieri and Diamond would both agree that, "apes became humans only when culture became their top competitive advantage", but also that "with culture, humans became the most dangerous creatures in the universe" (Ghiglieri, 1999, p. 78). What is more, both Ghiglieri and Diamond refer only to the common chimpanzee and ignore findings regarding the bonobo, known already at the time they conducted their studies. I would argue that it is not a coincidence. The reason is one and the same: the unconscious adoption of the teleological approach in anthropology. From the moment that Auschwitz and Hiroshima occurred, it is as if it could not have been otherwise. All that remains is to prove it on a socio-genetic level.

The Bonobo

> *Everything that ought to have remained secret and hidden but has come to light.* (Freud, 1919, p. 223)

The bonobo is a pygmy chimpanzee, genetically similar to the common chimpanzee and, as recent studies have shown, at least as closely related to us as the chimp. While the common chimpanzee, particularly the male, is a violent, murderous, meat-eater, the bonobo adheres to the Woodstockian

Fig. 3.2 A female bonobo copulating with another female bonobo. Credit Zanna Clay
https://cdn.mos.cms.futurecdn.net/hUzbw7jq6MwaPNNMoZkMdM-970-80.jpg.webp

ideal of "free love", engaging in sexual behavior frequently and in countless positions—males with females, females with females, and males with males. Sex (Fig. 3.2), for the bonobo, is a means of conflict resolution. Indeed, there is hardly any violence among bonobos, infanticide is non-existent, and unresolved tensions are rare. If everyone mates with everyone and paternity is unknowable, it would be rather stupid (evolutionally speaking) to kill an infant that carries half of your own genes.[3] No less important is the fact that it is not the males, but the females who are dominant (Fig. 3.3). Ironically, and contrary to the generally accepted comparison between humans and chimpanzees, bonobos spend a good part of their time in the savannah, walking on two, while chimps remain primarily in the trees.

[3] This argument accepts the idea of the selfish gene.

Fig. 3.3 Bonobos grooming each other

It is not the aim of this book to provide an exhaustive account of the state of evolutionary research, but to pose an important question regarding that research: why is it that scholars like Diamond, Ghiglieri, and others, in their attempts to understand our future by returning to our origins, have ignored the bonobo? The explanation—one with which the reader should be acquainted by now—seems simple enough: bonobo behavior does not fit the killer ape narrative, used to explain the killing fields of the twentieth century.[4]

The fact that Diamond ignores the bonobo is less significant, as his thesis is not based on the common chimpanzee. Ghiglieri, on the other hand, like anthropologist and primatologist Richard Ringham, explicitly relies on comparisons to the male chimp, arguing that human behavior follows a similar pattern to that of the common chimpanzee male, who

[4] We can have a secondary discussion of whether the bonobo has been ignored intentionally or unconsciously. In the studies I have seen, I have found no justification for ignoring the bonobo. For the purposes of the present discussion let us assume that it is not intentional—which does not make the fact that the bonobo has been ignored any less problematic, but at least precludes the suggestion of outright fraud.

attacks males from rival troops in order to gain control of their resources and females: "War is typically men's ultimate reproductive gamble. Many researchers agree that the goal most worth the lethal risk of war is women or the resources that attract or support more women and their offspring" (Ghiglieri, 1999, p. 197). According to Ghiglieri, both war and the extermination of all who are not "us" are a part of human nature. Ghiglieri argues that genocide is the result of "two human instincts ... ethnocentrism, focusing on one's own culture as the 'right' one, and xenophobia, fearing outsiders or strangers" (p. 211). Ghiglieri concludes, based on observations of male common chimpanzees, that these instincts are in our nature:

> First, field research on apes and humans has revealed to us that our own dark side is a deeply embedded set of instincts to kill, rape, or rob when it looks as if doing so might safely pay off reproductively or in our own survival. Second, much of human history is written around conflicts between "gene pools" that have decided, as a social group, to use these primitive and immensely selfish strategies to the hilt. (p. 217)

The fact that Ghiglieri ignores the bonobo is particularly disturbing in light of the reductive model he espouses, which views human history as a struggle between genes (we are merely survival machines—robotic instruments blindly programmed to preserve the selfish molecules known as genes), in the manner of Richard Dawkins. Ghiglieri also argues that humans and common chimpanzees share the traits of male retention and female out-migration. Genetic similarity is thus much greater between males than between females. According to Ghiglieri, studies of gatherer-hunter tribes have found a similar pattern. This appears to justify his association of human males with chimpanzees, while ignoring bonobos. Here, however, Ghiglieri's theory hits a snag, because "in nature, bonobo females do all the migrating, leaving their home group at puberty. Males stay put, enjoying the company and protection of their mothers" (de Waal, 2005, pp. 16–17). In other words, bonobos practice male retention and female migration, just like chimps. Nevertheless, Bonobo behavior differs significantly from that of common chimpanzees. At this point, it is hard to find any justification for Ghiglieri's ignoring the bonobo.

One could hardly argue for the existence of elements of bonobo-style female dominance among humans (although a number of tribes are said to have adopted such a model), yet the description of humans as identical to common chimpanzees is also far from reality (Wrangham, 2019). For example, the male-female ratio among bonobos is one to one, whereas the ratio among chimps is two females for every male, implying higher male mortality, probably due to internal strife—avoided in the bonobo political model, if only because males are not in constant competition for dominance over the group. The balance created within the bonobo group has obvious advantages, further strengthened by the absence of infanticide—apparently indicating a strong and stable strategy.

"The fundamental difference between our two closest relatives," writes Dutch primatologist and ethologist Frans de Waal, "is that one resolves sexual issues with power, while the other resolves power issues with sex" (2005, pp. 18–19). Despite their different behaviors, the extent to which bonobo and common chimpanzee instincts differ remains unclear. They are two very closely related species (species or subspecies?),[5] practically identical genetically, that present fundamentally different behavior

[5] There is a certain amount of diversity within every species. The question in this context is when do two individuals of the same species cease to be classified as such and come to be defined as distinct species? Is there an objective measure? The rule of thumb in such cases is that two individuals cease to be members of the same species when they are unable to produce fertile offspring. A mule, for example, is (generally) a cross between a male donkey and a female horse, but is, itself, sterile. As we shall see, this is a rather artificial definition, or, in the words of Darwin, "I was much struck how entirely vague and arbitrary is the distinction between species and varieties … The amount of difference considered necessary to give to two forms the rank of species is quite indefinite" (1998, p. 41–50). Darwin further writes: "Certainly no clear line of demarcation has as yet been drawn between species and sub-species—that is, the forms which in the opinion of some naturalists come very near to, but do not quite arrive at the rank of species; or … between sub-species and well-marked varieties … These differences blend into each other in an insensible series" (p. 44). Darwin even stresses that, "neither sterility nor fertility affords any clear distinction between species and varieties" (p. 201). Indeed, it often seems that "there is no species barrier" (Weiner, 1994, p. 39). Darwin therefore refers to varieties, when they are sufficiently well-marked, as "incipient species". Thus, "If a variety were to flourish so as to exceed in numbers the parent species, it would then rank as the species, and the species as the variety; or it might come to supplant and exterminate the parent species; or both might co-exist, and both rank as independent species" (1998, pp. 44–45). This is crucial, since we generally define the various species of *Homo* as distinct. Such definitions are retrospective, however, based on a discontinuous fossil record. In reality—that is, during the process of transformation—no such distinction between species could be made: every individual was more or less identical to its parent and its child. Dawkins (2009) thus argues that the notion of different species over the course of human evolution is purely conceptual and does not reflect the real world at any given time.

patterns. Perhaps this is a good time to ask what "natural instinct" is. De Waal writes, "The word 'instinct' stuck in my mind. I barely know what this means anymore, since purely inborn behavior is impossible to find" (p. 147).

Who Is the First Ancestor?

Let us now return to the comparison between the common chimpanzee and the bonobo, remembering that the two are equally related to us. According to de Waal (2005), the model adopted by humans is neither that of the common chimpanzee nor that of the bonobo: "Human social organization is characterized by a unique combination of (1) male bonding, (2) female bonding, and (3) nuclear families. We share the first with chimpanzees, the second with bonobos, and the third is ours alone" (p. 108). As we have seen, Ghiglieri focuses primarily on mating competition. In practice, however, the average human male can hardly be compared to the average male chimp or bonobo. Humans are both and a little more, and indeed, our mating model is different: "The intimate male-female relationship implied in all of this, which zoologists have dubbed a 'pair-bond,' is bred into our bones" (de Waal, 2005, p. 108). In human males this is reflected in smaller testicles (suggesting less mate competition), compared to those of our closest relatives: "As a proportion of body weight, our testicles are nearly five times as large as gorillas' and one-third the size of chimpanzees'" (Ridley, 2003, p. 21). De Waal goes as far as to claim that it is precisely this 'pair-bond' and paternal as well as maternal investment in offspring that led *Homo sapiens* to the great leap forward.

At this point I would ask, is it merely a coincidence that both Diamond and Ghiglieri depart from the chimpanzee and end with genocide? Put another way, have we really done away with the Aristotelian "final cause", that is, the idea that everything in nature acts for a particular reason and that is its purpose and its essence?[6] It is hard to ignore scientific findings regarding the common chimpanzee. For example, researchers in Gombe Stream National Park (Tanzania) observed a group of male chimpanzees

[6] Aristotle described four types of causes, which answer the following questions: (1) What is it made of? (material); (2) What is its essence? (formal); What caused it to come into being? (efficient); and (4) What is its purpose? (final). According to Aristotle, everything in nature has a purpose (telos, τέλος). For example, the telos of an acorn is to become an oak tree. As Bechler (1987) has shown, the scientific revolution largely did away with the final cause.

raised together in the same community later split into two separate communities. Despite their initial closeness, the chimps did not hesitate, when they met on the border between their respective territories, to murder one another and drink the blood of those they had killed. Although this description does evoke scenes of ethnic cleansing, it does not necessarily mean that "chimpanzee-like violence preceded and paved the way for human war, making modern humans the dazed survivors of a continuous, 5-million-year habit of lethal aggression", as anthropologist and primatologist Richard Wrangham (1996, p. 63) put it. At the same time, we cannot ignore findings regarding an encounter between two groups of bonobos in the Democratic Republic of the Congo. The groups did not fight or try to exterminate one another. Although there was some shrieking, chasing, and so forth, particularly among the males, there was no serious violence. Within 15 min, intergroup male-female and female-female sex was observed. The males remained somewhat hostile or indifferent to the males of the opposing camp, occasionally skirmishing, but even they "eventually engaged in brief scrotal rubbing" (de Waal, 2005, p. 139).

Note, I am not suggesting that the bonobo model is more appropriate to *Homo sapiens* than the chimpanzee model. When considering our origins, however, we should, at least, not ignore the bonobo—in the words of de Waal: "The last common ancestor of humans and apes may have been gorillalike, chimpanzeelike, bonobolike, or different from any living species" (p. 138).

Entrapped in the Chains of Memory of That Abyss

As we have seen, there is no real contradiction between studies that focus on "instincts" and those that focus on "culture", since the main purpose of the search for our origins as a species is to find an explanation for Auschwitz and Hiroshima. The (structural) trauma of Auschwitz leads us to rethink the evolutionary process. As the philosopher Adi Ophir put it:

> And all this is still entrapped in the chains of the memory of that abyss at the heart of the culture in which the Nazi nightmare is still alive and well, borne in the minds of those who were rescued and those who grew up in the shadow of the survivors, and of all the others inundated by the unending verbalization that sanctified the destruction and set up Auschwitz in the blank space of God. (2005, p. 592)

In the background is the premise that such nightmares are neither an accident nor an error but a deterministic, evolutionary, technological necessity—hence the primary reason for ignoring the bonobo. Indeed, since Auschwitz, "bonobos fail to fit established notions about human nature" (de Waal, 2005, p. 30).

To be clear, I am not suggesting that humans should be reduced to chimpanzees or bonobos. There is something of each in us, as well as subtle traits that are hard to pin down. In the words of British zoologist Matt Ridley: "Here is a mechanism for creating large and small evolutionary changes from small genetic differences. Merely by adjusting the sequence of a promoter, or adding a new one, you could alter the expression of a gene." What is more, "Just a *tiny change in one promoter will produce a cascade of differences for the organism.* These changes might be sufficient to create a wholly new species without changing the genes themselves at all" (Ridley, 2003, p. 33). These small changes may be what allow us to write, read, and worry about alien invasions, and especially, to be anxious of death and obsessed with the future. Some believe that this is what makes the difference between us and chimpanzees merely "one of degree" (p. 16).

It is also possible, however, that all discourse regarding the 1.6% genetic difference between humans and chimpanzees is simplistic demagoguery, since the very same issue can easily be approached from the opposite direction: "Genetically, the differences between a human being and a chimpanzee are at least 10 times as numerous as those between the two most dissimilar human beings" (p. 10). One might also argue that, "the difference between two individual human beings amounts, on average, to 0.1 percent, so there are 3 million different letters [of genetic code] between me and my neighbour. The difference between a human being and a chimpanzee is about 15 times as great, or 1.5 percent. That equates to 45 million different letters" (pp. 25–26). Employing other methods of calculation (replacements + additions + omissions without replacement), the gap increases to 5%. Conversely, some have argued that, "we differ by just a single atom on a humble sugar molecule, and an omission at that" (p. 27). These are empty words and hollow numbers—further testimony to the fact that we see the world from the perspective of technicians. It is a hopelessly simplistic way of thinking when, in reality, "the genes themselves will tell them little, and the source of human uniqueness will remain as mysterious as ever" (pp. 33–34).

It is not my contention that genetic research is irrelevant, just as I do not claim that everything is determined by our environment. What I am suggesting is that humans cannot simply be reduced to their genes and, more importantly in the context of this book, if we do choose to adopt a reductive approach, why limit ourselves to the violent male chimpanzee, while systematically and tendentiously ignoring bonobos? It would appear to be part of a broader tendency, which seeks to explain the horrors of the twentieth century. It's not our fault. We're just chimps with weapons of mass destruction.

References

Bechler, Z. (1987). The essence and soul of seventeenth-century scientific revolution. *Science in Context, 1*(1), 87–101.

Darwin, C. (1998). In G. Beer (Ed.), *The origin of species*. Oxford University Press.

Dawkins, R. (2009). *The greatest show on Earth: The evidence for evolution*. Bantam Press.

de Waal, F. (2005). *Our inner ape: A leading primatologist explains why we are who we are*. Riverhead Books.

Diamond, J. M. (1997). *Guns, germs, and steels*. W.W. Norton & Co.

Diamond, J. (2006). *The third chimpanzee: The evolution and future of the human animal*. Harper Perennial.

Ferguson, R. B. (2023). *Chimpanzees, war, and history: Are men born to kill?* Oxford University Press.

Freud, S. (1919). "The Uncanny." Standard Edition XVII: 217–253.

Ghiglieri, M. P. (1999). *The dark side of man: Tracing the origins of male violence*. Helix.

Ophir, A. (2005). *The Order of evils: Toward an ontology of morals* (R. Mazali, & H. Carel, Trans.). Zone Books.

Ridley, M. (2003). *Nature via nurture: Genes, experience, and what makes us human*. Fourth Estate.

Sussman, R. W. (2013). Why the legend of the killer ape never dies: The enduring power of cultural beliefs to distort our view of human nature. In D. P. Fry (Ed.), *War, peace, and human nature: The convergence of evolutionary and cultural views* (pp. 97–111). Oxford University Press.

Weiner, J. M. (1994). *The beak of the finch: A story of evolution in our time*. Knopf.

Wrangham, R. (2019). *The goodness paradox: The strange relationship between virtue and violence in human evolution*. Pantheon.

Wrangham, R., & Peterson, D. (1996). *Demonic males: Apes and the origins of human violence*. Mariner/Houghton Mifflin.

CHAPTER 4

What Happened to Neanderthal? (We Killed Him Off)

One of the burning questions in the history of *Homo sapiens* concerns its expansion from Africa to Europe. The reasons for the departure from Africa are unclear, but climate change—probably an ice age—is generally cited (Tierney et al., 2017). As a result of climate change, certain areas became more arid and less fit for human habitation, while others became more fertile and better suited to human needs. Some have suggested that *sapiens* simply followed the animals they hunted, unaware that they were migrating (Kübler, 2015). Others have posited that overpopulation, in and of itself, would have been reason enough. Milo goes as far as to say that "those who left Africa had no ecological reason to do so" (Milo, 2009, p. 96).

There were a number of waves of *Homo* migration from Africa to the Middle East and Europe. We know that *Homo erectus* left Africa and reached as far as China, 1.8 million years ago (Fig. 4.1).[1] As for *Homo sapiens*, the first wave of migration is generally believed to have occurred some two hundred thousand years ago, although scholars have recently suggested an earlier date, a date closer to three hundred thousand years ago (Fig. 4.2). This was followed by successive waves, in which tribes of *Homo sapiens* left Africa and, passing through the Middle East, reached as far as Europe. The difficulty in solving the riddle of human ascendance is that the process seems too rapid—as if there were some leap that turned

[1] See (Zhu et al., 2018).

Y. Ataria, *Genes, Technology, and Apocalypse*,
https://doi.org/10.1007/978-3-031-59197-6_4

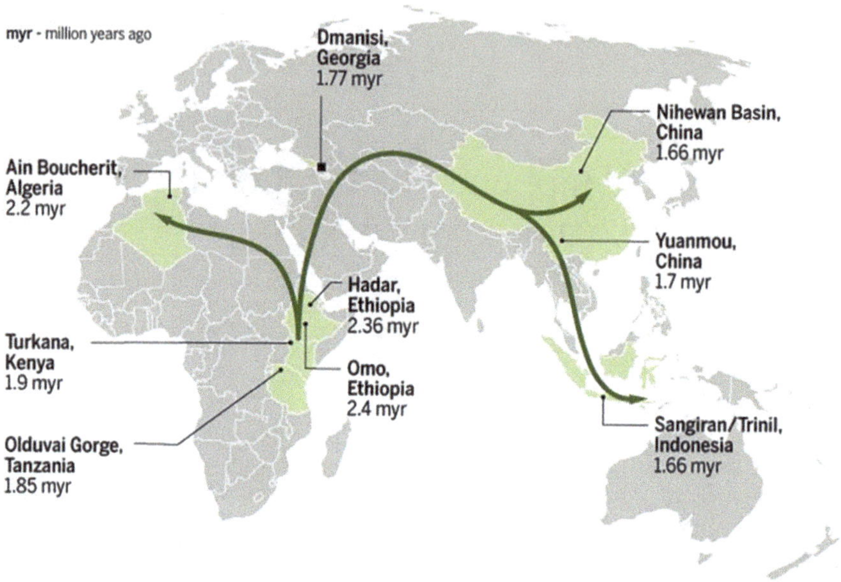

Fig. 4.1 Migration map—early Homo
By following a trail of stone tools and fossils, researchers have traced possible routes for the spread of early Homo out of Africa to the far corners of Asia, starting about two million years ago
https://becominghuman.org/wp-content/uploads/2019/10/bh-migration-map-science.jpg

Homo sapiens into an extraordinary species, capable of conquering all of Europe in under 20,000 years. Israeli ethologist and science writer Yonat Eshchar put it as follows:

> In the Middle East, for the most part, it seems that Neanderthals and modern humans lived side by side for tens of thousands of years. Elsewhere, particularly in Europe, there was hardly any overlap between the two populations. The earliest evidence of the presence of *Homo sapiens* in Europe is from 43,000–45,000 years ago—only a few thousand years, the blink of an eye in historical terms, before the disappearance of the Neanderthals. This temporal proximity makes us suspect a hostile takeover by these *Homo sapiens* of areas inhabited by Neanderthals, or even more direct conflict with them. (Eshchar, 2022)

As we have seen, scholarship has struggled to identify unique biological or cultural traits in modern humans that might explain *Homo sapiens*' takeover of the European continent (Fig. 4.2). Despite the difficulty in identifying such traits, the fact remains that we are indeed alone. Although we are unable to determine what exactly happened or why, it is generally agreed that, in the words of Diamond: "Cro-Magnon [*Homo sapiens*] arrival somehow caused Neanderthal extinction" (2006, pp. 51–52). Diamond believed that remains from eastern Europe and the Middle East, dated to between 60,000 and 90,000 years ago, indicate a prolonged struggle between the Neanderthals and the out-of-Africa *Homo sapiens*, and that the blitzkrieg known as the "great leap forward" had not yet occurred by then: "The slowness of the transition in the Near East, compared to its speed in western Europe, suggests that the anatomically modern people living around the Near East before sixty thousand years ago *had not yet developed the modern behavior* that ultimately let them drive out the Neanderthals" (p. 53). This assertion reflects the premise behind all of Diamond's work: what makes *Homo sapiens* unique (modern behavior) is the ability to make rival species disappear (expel the Neanderthals). With

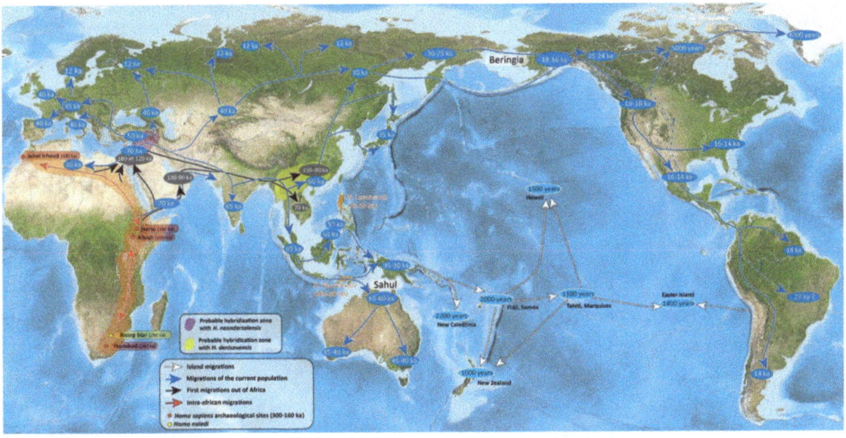

Fig. 4.2 *Homo sapiens* migrations and the modern population
Source: Sebastien Nomade (https://www.cea.fr/english/lists/staticfiles/clefs/science-history/images/partie1/migrations-homo-sapiens_original.jpg)

this in mind, I would describe Diamond as the leader of the third wave of what de Waal (2005) calls the "veneer theory". According to this theory, "Deep down, we humans are violent and amoral ... Our species never had the time to evolve the appropriate inhibitions. As a result, we don't have our fighting instinct under control as much as 'professional predators' like wolves or lions" (p. 20). Among the exponents of the first wave, we find Nazi Party member and Nobel laureate in physiology Konrad Lorenz, "the great defender of the idea that aggression is in our genes" (ibid., p. 21). The second wave, which "managed to outdo the first", includes British evolutionary biologist Richard Dawkins, author of *The Selfish Gene*, for whom "human decency is no more than a thin crust—something we invented rather than inherited" (ibid., p. 22); and Ardrey, who presented the "first ancestor" as "a carnivore who swallowed his prey alive, dismembering them limb from limb, to slake his thirst for blood" (ibid., p. 21). In the more sophisticated third wave we encounter Diamond, who departs not from the paleoanthropological evidence but from Auschwitz and Hiroshima, deducing backward to the male chimpanzee from which *Homo sapiens* evolved and went on to drive the Neanderthals from Europe.

As we have seen, the more scholarship advances, the smaller the gap between modern humans and Neanderthals becomes (Fig. 4.3). Both can be traced back to *Homo heidelbergensis* (or *Homo erectus*, broadly understood), who lived some 600,000 years ago (and it is worth pointing out, once again, that the division into two distinct species is by no means certain). The average Neanderthal brain was larger than that of *Homo sapiens* (1.2–1.7 L, on average, as compared to 1.4 L – see Fig. xx), indicative of high cognitive function, according to many scholars. The Neanderthals hunted big game, controlled fire, cooked, sewed clothing, created art, fashioned jewelry, practiced healing, and more. It is important to note that evidence of these activities predates their encounter with modern humans, so that the idea that they simply copied from *Homo sapiens* is unfounded. Furthermore, Neanderthals appear to have performed ceremonies and rituals, although no incontrovertible evidence of this has been found. All of the evidence points to a culture resembling that of *Homo sapiens*.[2]

[2] See (Roebroeks et al., 2012; Nielsen et al., 2020).

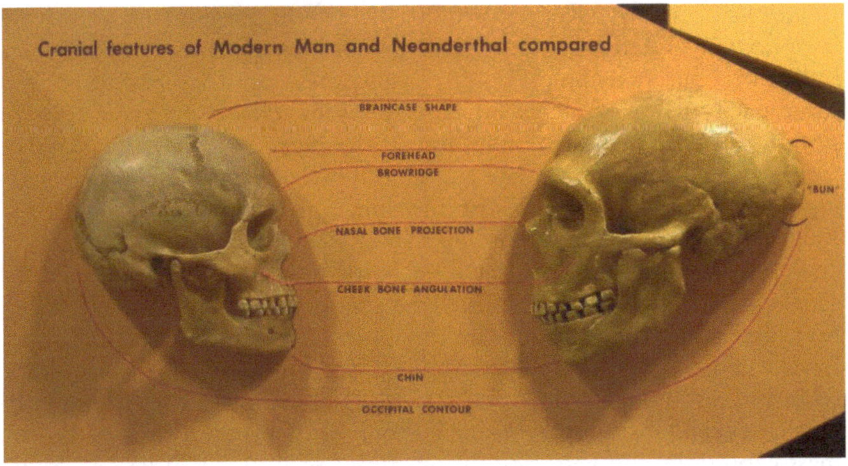

Fig. 4.3 *Sapiens* Vs. Neanderthal
Source: Comparison of Neanderthal and Modern human skulls from the Cleveland Museum of Natural History. https://upload.wikimedia.org/wikipedia/commons/thumb/f/f2/Sapiens_neanderthal_comparison.jpg/800px-Sapiens_neanderthal_comparison.jpg

Despite this, Diamond argues that the Neanderthals stood still: "Neanderthal tools from 100,000 and 40,000 years ago look essentially the same" (2006, p. 44). The focus here is on *innovation*, showing, once again, how scholars like Diamond seem incapable of relating to the findings themselves without the prism of the modern-day consumerist obsession with everything "new and improved". In other words, the findings themselves may be reliable, but the narrative is biased toward the present.

As we have seen in previous chapters, there is widespread agreement that the uniqueness of *Homo sapiens* is related to language. Recent genetic and anatomical research (the presence of the FOXP2 [Forkhead box protein P2] gene and the physiological structure that makes language possible) has shown, however, that *Homo neanderthalensis* also had the capacity to produce language (Rutherford, 2018).[3] It is important to note that we have no way of determining whether or not Neanderthals actually spoke, but from a purely technical point of view, they appear to have possessed the capacity to do so. This similarity between *Homo sapiens* and *Homo neanderthalensis* makes the question of the Neanderthals' disappearance from regions in which they had a clear, physical advantage (Fig. xx)—having developed in the cold European climate, while *sapiens* developed in the heat of Africa—all the more disturbing.

When Diamond observes the speed with which *Homo sapiens* replaced *Homo neanderthalensis* and immediately suspects that *sapiens* was somehow involved in the Neanderthals' disappearance, he commits two errors. First, when he chooses to describe this process in terms of driving off, he relies not on findings from 40,000 years ago—although the type of findings that could even support such a theory is by no means clear—but on mass expulsions and genocides in our time. Second, when he focuses on rapid technological change and the tendency to innovation, he is, once again, deeply influenced by the present age, in which barely a day goes by without hearing about some new and "revolutionary" gadget.

Essentially, any study of the great leap forward is, at the same time, also a study of the Neanderthals that once were but are no longer. Was it a matter of competition over resources? Was it a series of wars ultimately lost to modern humans? Or were the Neanderthals simply assimilated? We

[3] At the beginning of the twenty-first century, many scholars believed that changes in the FOXP2 gene (the first gene identified as pertaining to language, based on genetic analyses of a speech-impaired family) occurred some 200,000 years ago—corresponding more or less to the emergence of *Homo sapiens*. A gene mutated and the mystery is solved. The story of the "language gene" FOXP2 was published in 2002. The theory presented in this study was that if a mutation affords a significant evolutionary advantage, it will spread more quickly—a phenomenon known as a "selective sweep". But can language really be reduced to a single genetic mutation, as if the story of human evolution had been scripted by Hollywood? Indeed, research has shown this reductive description to be an oversimplification. To the extent that there is a connection between language and genes, it is a rather complex one, involving an entire network of genes that has changed over time. In any event, recent studies have shown that our relatives the Neanderthals carried the very same mutation of the FOXP2 gene initially believed to be unique to humans. We know that Neanderthal lineage diverged from our own around half a million years ago—indicating that the FOXP2 gene mutation in question occurred hundreds of thousands of years earlier than scholars had previously thought. Moreover, according to an extensive study, no selective sweep—that is to say an evolutionary change that conferred an exceptional advantage for survival—has occurred in the past 200,000 years.

now know that the two species did, in fact, interbreed.[4] The greatest proof of this is that nearly all non-African humans today carry a small amount of Neanderthal DNA (usually between 1 and 5%). Genetic studies have shown that this interbreeding occurred some 50,000–60,000 years ago, probably in the Middle East.

In all likelihood, it was all three—combined with a genetic weakness stemming from a relatively small gene pool,[5] making them more vulnerable to the diseases that modern humans brought with them from Africa (where a wide variety of bacteria, viruses, and parasites thrive even today) and had evolved to resist—not unlike the course of events that followed the European invasion of the Americas: "When European colonists invaded North America, most North American Indians proceeded to die of introduced epidemics; most of the survivors were killed outright or driven off their land" (Diamond, 2006, p. 52).[6]

In light of the above, I would like to return to Ardrey, who, in his eloquent and powerful but non-scientific writing, represents the post-Auschwitz consensus that we are born killers: "What happened to Neanderthal," he asks and answers rather bluntly: "I find no mystery at all. We killed him off" (1976, p. 167). Of course, this claim cannot be proved and is, as we have seen, almost certainly inaccurate. It seems almost as if *Homo sapiens* invaded Europe in a well-planned blitzkrieg. In this answer, however, as paleoanthropologist Clive Finlayson (2009) has suggested, we find redemption, exoneration. Auschwitz, after all, is not our fault. We were always like that. We were killers from the start—one on one, at first, and then systematically: "The famous *Homo habilis* of two million years ago apparently died of a blow on the top of his head" (Ardrey, 1976, p. 163). Humans are thus not to blame; it is who we are—inveterate killers who, in this day and age, kill under the aegis of bureaucracy.

[4] Although the possibility of sex between modern humans and Neanderthals was barely considered until a few years ago, recent genetic studies leave little room for doubt. For example, a 50,000-year-old Neanderthal female discovered in 2014 bears genetic traces indicative of interbreeding between modern humans and Neanderthals some 50,000 years earlier (that is, 100,000 years ago). Furthermore, such interbreeding is confirmed by our own genetic makeup, particularly among Europeans (Rutherford, 2018). It is worth noting here that the fact that that sex between modern humans and Neanderthals produced fertile offspring precludes the argument that we belong to distinct species—at least according to the simple definition of the concept.

[5] See (Vaesen et al., 2019).

[6] See (Greenbaum et al., 2019).

This is the story we like to tell today: evolutionary change gave rise to cultural change, which, in turn, influenced our biology. For example, weapons and fire eliminated the need for large canines, but when they shrunk, we became all the more dependent on weapons and other tools. Dependence on technology created a slippery slope: "The biological changes that came about in consequence of cultural advance forbade retreat" (Ardrey, 1976, p. 138). In the process, we became a culturally dependent creature trapped in a relatively weak biological body that requires a complex brain in order to enable extensive cooperation. This is both the trap and the explanation: a hunter who has lost significant physical advantages in exchange for cultural development. We are thus little more than "prisoners of cultural advances" (p. 134).

References

Ardrey, R. (1976). *The hunting hypothesis: A personal conclusion concerning the evolutionary nature of man*. Atheneum.

de Waal, F. (2005). *Our inner ape: A leading primatologist explains why we are who we are*. Riverhead Books.

Diamond, J. (2006). *The third chimpanzee: The evolution and future of the human animal*. Harper Perennial.

Eshchar, Y. (2022, February 21). *When did we come to Europe: Earlier than we thought (Heb.)*. Retrieved from Davidson Institute of Science Education: Science News: https://davidson.weizmann.ac.il/online/sciencenews/

Finlayson, C. (2009). *The humans who went extinct: Why neanderthals died out and we survived*. Oxford University Press.

Greenbaum, G., Getz, W. M., Rosenberg, N. A., Feldman, M. W., Hovers, E., & Kolodny, O. (2019). Disease transmission and introgression can explain the long-lasting contact zone of modern humans and Neanderthals. *Nature Communications, 10*, Article 5003. https://doi.org/10.1038/s41467-019-12862-7

Kübler, S. e. (2015). Animal movements in the Kenya rift and evidence for the earliest ambush hunting by hominins. *Scientific Reports, 5*, Article 14011. https://doi.org/10.1038/srep14011

Milo, D. S. (2009). *The invention of tomorrow (Heb.)*. Hakibbutz Hameuchad.

Naama, G.-I. (2011). Behavioral and cultural origins of Neanderthals: A Levantine perspective. *Continuity and Discontinuity in the Peopling of Europe: One Hundred fifty Years of Neanderthal Study*, 89–100.

Nielsen, M., et al. (2020). Homo neanderthalensis and the evolutionary origins of ritual in homo sapiens. *Philosophical Transactions of the Royal Society of London. Series B, Biological Sciences, 375*(1805), Article 20190424. https://doi.org/10.1098/rstb.2019.0424

Roebroeks, W., et al. (2012). Use of red ochre by early neandertals. *PNAS, 109*(6), 1889–1894. https://doi.org/10.1073/pnas.111226110

Rutherford, A. (2018). *A brief history of everyone who ever lived*. Weidenfeld & Nicolson.

Thomas, W., Frederick, L. C., & Fred, C. (2011). *How to think like a Neandertal*. Oxford University Press.

Tierney, J. E., deMenocal, P. B., & Zander, P. D. (2017). A climatic context for the out-of-Africa migration. *Geology, 45*(11), 1023–1026.

Vaesen, K., et al. (2019). Inbreeding, allee effects and stochasticity might be sufficient to account for neanderthal extinction. *PLoS ONE, 14*(11), Article e0225117. https://doi.org/10.1371/journal.pone.0225117

Zhu, Z., Dennell, R., Huang, W., Wu, Y., Qiu, S., Yang, S., ... & Ouyang, T. (2018). Hominin occupation of the Chinese Loess Plateau since about 2.1 million years ago. *Nature, 559*(7715), 608–612.

CHAPTER 5

Collapse of the First Line of Defense: It's Not Us; It's in Our Genes

If humans are simply killer apes lurking behind suits and PhDs (Fig. 5.1), the road to Auschwitz and Hiroshima is open and clear. There is an evolutionary explanation of human actions.

According to evolutionary theory, species are constantly changing. What drives evolution is the mechanism of natural selection, which favors the survival and reproduction of individuals best suited to their environment (not necessarily the strongest). This is the engine behind evolutionary change and the source of diversity and variety in nature. Natural selection is the direct result of an imbalance between the natural fertility of an organism and the capacity of the environment to support it, with the first being greater than the second, meaning that some will not survive. "Thus extinction and natural selection will, as we have seen, go hand in hand … *The extinction of species and of whole groups of species … almost inevitably follows on the principle of natural selection*" (Darwin, 1998, pp. 141, 384). It is hard to overestimate the importance of this idea and the ways in which it has shaped social, political, and cultural structures over the past 200 years.

Dawkins, one of the leading advocates of the theory of evolution, argues forcefully that the struggle that Darwin refers to occurs on the level of the single gene and that humans are merely the sum total of their genes: "We, and all other animals, are machines created by our genes" and "a predominant quality to be expected in a successful gene is ruthless selfishness" (2006, p. 3). According to this reductive theory, humans and all other organisms are merely gene carriers: "A body is the genes' way of

Y. Ataria, *Genes, Technology, and Apocalypse*,
https://doi.org/10.1007/978-3-031-59197-6_5

preserving the genes unaltered" (ibid., p. 29). If we are slaves to the selfish gene, then Auschwitz and Hiroshima are easily explained. When the restrictions that humans have imposed on themselves—culture, law, penal justice, and so forth—are somehow lifted, death camps will result. According to this approach, Auschwitz taught us the true meaning of ruthless selfishness. In other words, Auschwitz is a point in time and space in which "human nature" is fully revealed:

> Natural selection favours genes that control their survival machines in such a way that they make the best use of their environment. This includes making the best use of other survival machines, both of the same and of different species ... The logical policy for a survival machine might therefore seem to be to murder its rivals, and then, preferably, to eat them. (ibid., pp. 86–87)

Without getting into specific intra-disciplinary scientific discourse, the selfish gene approach—although it does provide a number of elegant explanations (for the phenomenon of altruism, for example)[1]—clearly

[1] According to the gene-centered approach to evolutionary biology (which views the gene as the fundamental unit of selection), genetic relatedness between individuals explains, for example, why songbirds and other animals may warn fellow flock or herd members at the cost of placing themselves in greater danger (although it is worth noting that such cries are, in fact, very hard to locate). Such behavior seems rather odd from an evolutionary perspective, since the individual organism lies at the heart of Darwinian Theory: "Altruism was a puzzle ... blatantly opposed to the fundamentals of the theory" (Harman, 2011, p. 3). If a given individual belongs to a flock, not only do its own genes have a greater chance of survival thanks to the protection offered by the flock, but its genetic similarity to other members of the flock could mean that it is worth the selfish gene's while to protect the entire flock: "Bird alarm calls have been held up so many times as 'awkward' for the Darwinian theory that it has become a kind of sport to dream up explanations for them. As a result, we now have so many good explanations that it is hard to remember what all the fuss was about. Obviously, if there is a chance that the flock contains some close relatives, a gene for giving an alarm call can prosper in the gene pool because it has a good chance of being in the bodies of some of the individuals saved ... Once again, uttering a warning call is seen to have a purely selfish advantage" (Dawkins, 2006, pp. 219–221). Solutions of this kind (genetic relatedness) also explain the "altruistic suicide" of bees and other fascinating phenomena. Ultimately, however, they reduce us to little more than "shrewd genetic calculators" (Harman, 2011, p. 161). Thus—despite a number of unresolved issues, such as the mechanism by which an individual recognizes genetic kinship with other individuals—"The purity of selflessness is undermined by the scourge of self-interest: What looks like sacrifice may in fact be the road to personal gain" (p. 5). Note that, "biologists consider a trait altruistic when it decreases the relative fitness of the individual that bears it, while increasing the relative fitness of other members of the group" (Shavit, 2008, p. 7). Indeed, when we look at evolution on a genetic level, what appears to be altruistic is in fact "for the good of the gene": "Both individual selfishness and individual altruism are explained by the fundamental law that I am calling gene selfishness" (Dawkins, 2006, p. 8).

raises some difficult questions. First, Dawkins struggles to define the unit "gene"; secondly, if we accept Dawkins's definition, if only as a metaphor, there is evidence that the mechanism of natural selection acts at higher and more complex levels (from genetic networks to entire communities); and thirdly, it is not at all clear that Dawkins has solid evidence to support the claim that behavior can be explained by a single unit of any kind:

> What has been discovered is a correlation between the presence of a particular DNA sequence and the presence of the character. Usually it is not at all clear that the DNA sequence is *causally* related to the character, and it is almost always very clear that "the gene" is neither a sufficient nor a necessary condition for the character's development. (Jablonka & Lamb, 2005, p. 59)

In my opinion, however, Dawkins has a far more serious problem. Like Ardrey, he is motivated by the need to explain how we got to Auschwitz, or, more precisely, how to exonerate us and relieve us of responsibility for acts of genocide. The selfish gene offers such an explanation, albeit in a more sophisticated manner than Ardrey's killer ape hypothesis, as it allows us to exclude all non-professionals from the debate, so that any natural phenomenon can be explained, with the help of a suitable mathematical formula, from the perspective of the selfish gene (it may look like kindness, but it is actually brutal competition and mutual exploitation). At the same time, the selfish gene theory tells us exactly what we want to hear: *humans are not to blame, because it is literally in our genes.* By this point, it should be clear that the two theories are, in fact, identical.

Thus far, we have explained the tendency to search for Auschwitz in the origins of our species (the quest for the "first human"), as well as the reasons for this search. As we have seen, however, the evolutionary theories that reduce humans to killer apes or ruthless survival machines present scientific difficulties that force us to seek other explanations for the horrors of the twentieth century. This led to the idea that we are not in fact slaves of our genes, but of technology. And voila, we have an alternative explanation for Auschwitz: it's not in our genes; it's technology's fault! In the next chapter, we will discuss the concept of "technological autonomy" and see how the narrative of the techno-scientific dynamic can explain the horrors of the twentieth century, and, more importantly, absolve us of responsibility for them.

References

Darwin, C. (1998). In G. Beer (Ed.), *The origin of species*. Oxford University Press.

Dawkins, R. (2006). *The selfish gene*. Oxford University Press.

Harman, O. (2011). *The Price of altruism: George Price and the search for the origins of kindness*. W. W. Norton & Company.

Jablonka, E., & Lamb, M. J. (2005). *Evolution in four dimensions: Genetic, epigenetic, Behavioral, and symbolic variation in the history of life*. MIT Press.

Shavit, A. (2008). *One for All? Facts and Values in the Debates Over the Evolution of Altruism (Heb.)*. Magnes Press.

PART II

It's Technology's Fault

The attempt to reduce humans to killer apes or blood-soaked genetic machines presents a number of fundamental difficulties, including the following: (1) There is no good reason to ignore the bonobo and prefer the violent male chimpanzee as a reference point for human evolution; (2) Just as there are violent gatherer-hunters, there are non-violent hunter-gatherers, making it very hard to sustain the hunting myth; (3) Assuming that *Homo sapiens* and *Homo neanderthalensis* evolved from a common ancestor, there is no reason to believe that *sapiens* was more "malevolent" or cruel and therefore survived at the expense of the Neanderthals—who, in fact, survived in a particularly hostile environment; (4) It is hard to ignore the fact that the various models that portray humans as bloodthirsty and violent are male-centered and chauvinistic to the point that they ignores the role of women in the evolution of the species. Since it is clear that women did, in fact, play an important role, many of the theories that focus on the violent male would appear to be entirely without basis.

Nevertheless, popular culture continues to sustain this narrative (the hunting myth and all it entails). Most people have never even heard of the bonobo, nor are they aware of the methodological problems with Steven Pinker's studies—both in the way in which he selects the tribes on which he focuses as well as the way in which he chooses to interpret the findings. They are equally unfamiliar with the fact that the Neanderthal population in Europe was relatively small (numbering in the thousands or tens of thousands) and, what is more, so widely scattered that genetic exchange was very low, weakening the population as a whole and making it more susceptible to disease. Those familiar with the actual science, however, know full well that that if it is exoneration of the human species we are after, we will have to look elsewhere.

The first line of defense, "It's not us; it's in our genes", clearly does not hold up to scientific scrutiny. The question of how to absolve humanity of its actions thus becomes more difficult and pressing than ever—leading to the next line of defense: "It's technology's fault." If technology is an autonomous system that also controls other systems (economic, social, etc.) and ultimately humanity itself, the horrors of the twentieth century can certainly be laid at its door.

When we say that the technological system is autonomous, we mean that it acts within an internal dynamic and an internal logic, to achieve ends that are not dictated from without. According to French Christian philosopher, sociologist, and theologian Jacques Ellul, best known for his *The Technological Society* (1964), it is not merely that **t**echnology, as part of a socio-political-economic system, influences the development of humanity and human society, but that it is more fundamental than the other systems: it is not part of them, but above them. The primacy of the technological system stems from the application of techno-logic (and by that I mean the logos of the techno) to all areas of life. It is the principle of *total efficiency* that dominates all human endeavors.

The two central figures in the discourse about technology are Prussian-born philosopher and father of modern socialism Karl Marx (1818–1883) and Nazi philosopher Martin Heidegger (1889–1976). Marx was the first to point out that technological systems tend to follow an internal dynamic—within themselves and to serve their own ends. As he argued, however, in *The Poverty of Philosophy* (Marx, 1955), **t**echnology is controlled by the upper classes. Marx thus viewed the technological dynamic through the lens of his socio-economic analysis, in the context of the eternal struggle between oppressor and oppressed and the need for equitable distribution of the means of production. He refused to recognize the primacy of the technological system over other systems, because he believed that technologies are only significant inasmuch as they are part of the means of production and thus contingent on the social and economic systems that determine public and individual priorities.

Heidegger, on the other hand, ascribed primacy to technology over all other systems. He believed that **t**echnology is the fundamental state of being in Western culture. When I write **t**echnology with a highlighted "t", that is what I am referring to. But when did this state of being first emerge?

Is it a modern invention? Before we can relate to these questions and approach the complex thought of Heidegger, we must first understand that Heidegger viewed technology as having a certain essence. This essence, worth understanding at the outset, flows from the depths of traditional philosophical thought and has characterized metaphysics since the days of Plato and Aristotle.[1] Thus, even if it is the case that technology manifests itself in the present era in its most radical form, the philosophical foundations and premises that led it to function as a complete practical and theoretical system, foundational to the human experience, lie deep within the history of Western thought. We will return to this complex topic later on, as we try to understand it as fully as possible.

We will approach the Heideggerian concept of "essence" (Ger. *Wesen*) through the example of the essence of technology which, he asserts, "is by no means anything technological" (Heidegger, 1977, p. 4). According to Heidegger, technology is not a thing or a tool, but a total system, an order into which we are thrown and within which we act and are acted upon. Technology is a way of being—not an object to be grasped, but what shapes the way in which we grasp the world and ourselves. Technology places nature and humans within a framework that precludes thought in terms other than those of maximum exploitation.

Combining Marxist thought (an internal dynamic within the framework of historical determinism) and Heideggerian thought (metaphysical primacy) allows us to take a deterministic approach to the development of technology.[2] The question that immediately arises is who controls the system? Ellul would argue, following Heidegger, that the question itself is fundamentally misplaced—further evidence of the fact that we do not really understand the meaning of technology. According to Ellul, the thing we currently call "rationality" is none other than technological thinking—hence its primacy over all other systems. Heidegger and Ellul claim that technology should be treated as a system operating within the framework of the laws of efficiency—that is, technological rationality. The

[1] Metaphysics (Gk. μετά, beyond + φυσικός, natural) is the branch of philosophy that deals with the essence of things, exploring such questions as what is cause and effect, what is the nature of matter, why is there something and nothing— questions that would remain open for discussion even if all scientific questions were solved.

[2] There may be no need for such a synthesis, since Marx's arguments on the subject of technology are ultimately untenable, that is to say, technological catastrophe is inherent to Marxism, which essentially offers a secular eschatology.

aspiration to total efficiency, which resembles a natural force in a technological society, shapes the entire world, including its various systems.

Technology, as noted, is not something that can be grasped or controlled, but a particular way of thinking—a world view. When we try to quantify the world in Excel tables, when we think we can recreate primordial natural processes in the laboratory, when we believe that computer models can explain when species will cooperate and when they will prefer to compete, when we ask ChatGPT questions about the essence of life and expect meaningful answers, we are living within the technological state of being. In any event, reality cannot be transformed into Excel tables, competition between species cannot really be predicted by a computer model, and ChatGPT will never give meaning to our lives. Living within the technological state means believing in technology, as we do today. More than believing in it, however, it means embracing the technological-engineering way of thinking (the result of which is the banality of evil). The Nazi Final Solution, for example, was the result of transforming an essential question regarding the place of the Jews in European culture into a technological-engineering one. The more we rely on technology, not as a tool but as a way of thinking, to resolve essential problems in our lives, the more engineering and functional the solutions we arrive at will be—as if there were no room left for humanity.

Ellul did not think of technology only in terms of an all-encompassing state, but also as an actual system—like economic or social systems. He viewed technology as an autonomous system—that is, one that acts to realize goals it sets for itself, without practical or theoretical human intervention. Some reject the idea of technological autonomy, arguing that technology is flexible and mutable, and subject to the influence of other systems. As Ellul explained, however, the fact that technology is flexible does not make it any less autonomous—on the contrary. The flexibility of the technological system and its ability to integrate with other systems—economic, educational, political, ecological, medical, industrial, military, sexual, scientific, ethical, and other—while subordinating them to technological knowledge, make it a unifying and synchronizing factor. Put another way, the various other systems already operate within the technological field (like a field of gravity).[3] The technological system is, in fact,

[3] The technological society is not simply the result of rapid technological change. Winner (1986) shows, for example, that a technological society failed to emerge in Europe in the early second millennium, despite rapid technological change. Technology occupied a different

the inescapable environment, climate, atmosphere, texture, fabric, space, and field within which we operate.

Whether consciously or unconsciously, those who maintain that technology is autonomous often espouse another, more extreme position: that technology is a deterministic system.[4] The one does not necessarily follow

place in individual and collective consciousness at the time. It did not shape public or personal agendas, and people could envision themselves as separate from technological being. The technological society is the product of a particular mood. It is a state of mind in which it is impossible to conceive of reality outside of technological being. Until the eighteenth century, Ellul argued, craftsmen sought to improve tools in order to make it easier to perform the tasks associated with them, but technological improvement was not a goal in and of itself. Tools were merely means to an end. Up to that point, societies were characterized by local, traditional techniques, developed within their respective environmental, historical, cultural, and political contexts. For hundreds and thousands of years, knowledge was transmitted vertically (intergenerationally), progress was slow, and technology was shaped by local, cultural conditions. It is thus understandable why societies rejected technologies inconsistent with local culture and its value and belief systems. Innovation was not sanctified and change was accepted in dialogue with a variety of systems and traditions, and although efficiency was important, it was not yet the determinant factor—resulting in a broad range of manufacturing methods. According to Ellul, technology became an important element in our lives over the course of the eighteenth and nineteenth centuries. He saw the French Revolution and the process of Enlightenment as expressions of the Technological Revolution internalized by the bourgeoisie and entrenched in a system of practices and beliefs: enlightenment, scientific progress, liberalism, human rights, technological developments, rationality, humanism—all perceived as a single vector. In Ellul's opinion (Ellul, 1980), the technological society embraces the principle of total efficiency—not merely in the sense of trying to get the most out of a system of machines, or seeking to maximize profits, but as a standard in every aspect of life. In the technological society, the logic of techno serves as the moral foundation not only for bureaucracies but also for individuals: technological thought becomes the value system on the basis of which everything is judged, and any attempt to act outside the technological framework constitutes a paradigmatic change.

[4] The concepts of "autonomy" and "determinism" can be thought of in a number of different ways: autonomy is a condition (necessary but not sufficient, necessary and sufficient) for determinism; determinism is a condition (necessary but not sufficient, necessary and sufficient) for autonomy; autonomy and determinism are independent of one another. The definition of each of these concepts has a profound impact on the nature of the connection between them. In recent decades, serious efforts have been made to eradicate the claim that technology is deterministic. Without getting into the philosophical discussion, it seems that current technological developments are actually beginning to reveal the essence of technology—to render the element in which we live technological. The technological system is thus much more than deterministic; it is the element itself. Moreover, we cannot deny that technologies such as the television, the computer, and others have fundamentally transformed humankind. Indeed, many studies point to the deep connection between technological developments and historical shifts (Smith & Marx, 1994). Even if, from a purely logical

from the other, but once we have accepted the idea that technology is an autonomous system, little effort is required to espouse the more radical view: that the technological system is both autonomous and deterministic. Those who argue that technology is deterministic are not suggesting that the world can be reduced to technology (like the reduction of nature to the physical elements), but that the technological system is inherently autonomous and, consequently, limits human freedom of action. Ellul viewed human freedom and technological autonomy as opposing vectors. Technology creates an illusion of choice, while, in reality, limiting choice to the point of eliminating it entirely. With every new technology we embrace, the environment in which we operate changes, the field of action becomes more technological, and every attempt to retreat, even a little, becomes a zero sum game.

* * *

The first chapter in this second part of the book begins with an introduction to one of the twentieth century's most important philosophers: Martin Heidegger. The first section of the chapter discusses the attitude of Heidegger "the man" and Heidegger "the philosopher" to the Nazi Party, Hitler, and anti-Semitism, including questions such as whether Heidegger should be considered a Nazi philosopher, and what kind of anti-Semitism may be ascribed to him. Heidegger was undoubtedly the thinker who exerted the greatest influence on the philosophy of technology in the second half of the twentieth century.[5] In the section of this chapter, I will present Heidegger's position on the essence of technology, not as a tool, but as a total reality. The third section of the chapter will explore the

perspective, technology cannot be proven to be deterministic, because there will always be counterexamples, from the very fact that so much of popular culture relates to it as if it were, we, as a society, submit to and efface ourselves before it, making it more deterministic than it actually is. In a profound sense, even the attempt to discuss technology in terms of determinism and autonomy, limits discourse. Water does not cause fish and air does not cause mammals. They are merely enabling ecological systems. Nor does water determine whether there will be nasty sharks or friendly dolphins, but is the element within which various niches develop. The analogy to technology is somewhat expansive and vague, but the more frequent technological developments become, the more the ecosystem itself changes and technology determines how we will be thrown into the world. Technology is not a vector that negates human freedom. It is the element that defines the realm of possibilities within which we are able to contemplate the questions of choice, time, life, and humanity.

[5] For wider debate see: (Ataria & Tamir, 2023).

connection between Heidegger's philosophy of technology and Nazism, and suggest that his famous remark, "Agriculture is now a motorized food industry, the same thing in its essence as the production of corpses in the gas chambers and the extermination camps", was not incidental but inherent to his view on the essence of technology. In the final section of the first chapter, we will see how the theory that Heidegger developed regarding the essence of technology allowed him to completely exonerate the Nazi regime of the horrors it perpetrated and even to blame European Jewry for its own destruction.

The second chapter in this part of the book will discuss the scientific dynamic that led to the development of the atomic bomb—widely considered, in both scientific and popular literature, the culmination of the techno-scientific dynamic. In the first section of this chapter, I will describe the roles of leading scientists—Niels Bohr, Lise Meitner, Leo Szilard, Hans Bethe, Enrico Fermi, Albert Einstein, Robert Oppenheimer, Werner Heisenberg (who worked for the Nazis), and others—in the development of the bomb. In the second section of the chapter, we will take a closer look at the figure of Robert Oppenheimer, "father of the atomic bomb", who explained the development of the bomb in terms of "technical sweetness". Like Heidegger's explanation, this approach, to some extent, also absolves humankind of responsibility. Part II concludes with a discussion of the validity of humanity's second line of defense: "It's not our fault; it's technology".

References

Ataria, Y., & Tamir, L. (2023). Why overcoming Heideggerian intellectualism should precede overcoming metaphysics. *Human Studies*, 1–23.

Ellul, J. (1980). *The technological system*. (J. Neugroschel, Trans.). Continuum.

Heidegger, M. (1977). *The question concerning technology and other essays*. (W. Lovitt, Trans.). Harper & Row.

Smith, M. R., & Marx, L. (1994). In M. R. Smith, & L. Marx (Eds.), *Does technology drive history?* (pp. 1–36). MIT Press.

Winner, L. (1986). *The whale and the reactor: A search for limits in an age of high technology*. University of Chicago Press.

CHAPTER 6

Technology As Metaphysics: Heidegger's Attempt to Exonerate the Nazi Movement

Heidegger and Nazism

"The Führer Himself and He Alone Is German Reality and Its Law"

Heidegger, like Hitler, was born in 1889. His father was a church sexton and Heidegger was raised a Catholic and attended a Jesuit seminary. At the age of 20, he began to study theology and hoped to become a priest but never completed his training. In 1914, he obtained a doctorate in philosophy.

In 1919, Heidegger was appointed assistant to Jewish philosopher Edmund Husserl (known as the father of phenomenology) at the University of Freiburg, and in 1923, he accepted a professorship at the University of Marburg. In 1927, Heidegger published his seminal work, *Being and Time* (*Sein und Zeit*), which he dedicated to Husserl. In 1928, Husserl retired and Heidegger succeeded him in the chair of philosophy at the University of Freiburg.

In December 1932, Prof. Wilhelm von Möllendorff was elected rector at Freiburg, taking up the post in April 1933. Von Möllendorff rejected the ideology of the Hitlerian regime and refused to discriminate against Jewish faculty. Considerable pressure was brought to bear on him and, after only 5 days in the position, he stepped down. That same month,

Y. Ataria, *Genes, Technology, and Apocalypse*,
https://doi.org/10.1007/978-3-031-59197-6_6

Heidegger was elected to the rectorate in his place. Of 93 faculty members, two abstained. The 13 Jewish members of faculty did not take part in the vote, as they had been suspended a few days earlier (6 April 1933).

In March 1933, Heidegger joined the Cultural-Political Working Community of German University Professors (*Kulturpolitische Arbeitsgemeinschaft deutscher Hochschullehrer*)—an organization that sought to promote the Nazification of German academia. On 1 May 1933, he joined the Nazi Party, to which he continued to pay membership dues until 1945 (card number 312589, Gau Baden).

On 27 May 1933, Heidegger delivered his inaugural address as rector of the University of Freiburg, titled "The Self-Assertion of the German University".[1] He decided that the Nazi anthem (the *Horst-Wessel-Lied*) would be sung during the ceremony and the "*Sieg Heil*" salute would be given. Heidegger saw the Nazi salute not as "an avowal of Party membership ... [but] as an expression of the individual's ... inward identification with the new Germany" (Ott, 1994, p. 153).

As it often happened with Heidegger, many failed to understand his address. Even without understanding the address, however, there is no mistaking his use of words clearly identified not only with the Nazi ideology, but with the figure of Hitler: "All faculties of will and thought, all strengths of the heart and skills of the body, must be unfolded *through* battle [*Kampf*], heightened *in* battle, and preserved *as* battle" (Heidegger, 2003, p. 10 [emphasis in the original]).

In his rectorship address, Heidegger argued forcefully that academia must be bound to the destiny of the German nation. He characterized this bond as an imperative that takes precedence over scientific research and academic freedom. In the address, Heidegger also drew upon elements from his early philosophy, such as "Dasein" and "resoluteness", integrating them with concepts such as "power", "nation", and "leadership". In this context, it is worth noting the observations of the German philosopher Karl Löwith, who was a student of both Husserl and Heidegger. In a discussion on the political ramifications of Heideggerian thought, Löwith argued that there is a deep connection between authentic "existence" (one of the central ideas in *Being and Time*) and the concept of *Müssen* (duty) that follows from it, and the idea of "specifically German

[1] In some ways this was a watershed for Heidegger, leading some to suggest that he was, from this point on, "another Heidegger"—the later Heidegger. It is, of course, an arbitrary determination.

existence". This connection serves as a broad philosophical infrastructure within which authenticity leads to a range of other existential categories—"to decide for oneself", "wanting one's ownmost destiny", "to take stock of oneself in face of nothingness"—and ultimately to political involvement aimed at "destruction" (Löwith, 1993, p. 173). Löwith's use of the term destruction, in the sense of the "destruction of metaphysics", can hardly be a coincidence, since that is precisely the philosophical aspiration that Heidegger hoped to realize through political action.

To return to Heidegger's speech (Self-Assertion), he concluded the address as follows: "But we fully understand the splendor and the greatness of this setting out only when we carry within ourselves that profound and far-reaching thoughtfulness that gave ancient Greek wisdom the word: 'All that is great stands in the storm...' (Plato, *Republic*, 497d, 8)" (Heidegger, 2003, p. 11). Heidegger thus appears to have believed already in 1933 that the Nazi movement offered Western culture a new beginning—a way out of technological nihilism and meaninglessness, and a return to philosophy and poetry. It is also interesting to note that even at this stage, Heidegger identified the great danger inherent in Nazism, yet—and perhaps for this very reason—chose to join its ranks. On 21 August 1933, Heidegger adopted the *Führerprinzip*, thereby abolishing the election of the rector by the university senate and subordinated himself directly to the minister of education. Earlier that month, Heidegger had been appointed *Führer* of Freiburg University. On 3 November 1933, he issued a directive denying financial support to non-Aryan students and to students who did not belong to organizations such as the SS or SA. As *Führer-Rektor*, he dismissed a number of academics (Hermann Staudinger, Max Müller) whose positions, he felt, were not consistent with the party spirit.[2]

In July 1933, Heidegger defended five Jewish professors against dismissal, arguing that such a move would harm the reputation of German universities and should therefore be avoided, although the policy of dismissing Jews from academia was, in itself, worthy and correct. He also prevented the dissemination of anti-Semitic pamphlets. At the same time, however, there is no record of a single word of criticism on his part, regarding the establishment of a concentration camp near the university.

Heidegger supported not only the Nazi Party, but Hitler himself. In reply to an offer he had received from Munich University, for example, he wrote,

[2] In other cases (such as that of Eduard Baumgarten), he used his power to harm the careers of those who disagreed with his philosophical views, blocking their appointments at other universities.

"For me it is clear that, putting aside all personal motives, I must decide to accomplish the task that will allow me to best serve the work of Adolf Hitler" (Sheehan, 1988, p. 38). On a number of occasions, he also gave public voice to these views. For example, on 3 November 1933, three months after Hitler came to power, he published an article that concludes with the following words: "The Führer himself and he alone is German reality and its law, today and for the future" (ibid., p. 39). Nor were these statements detached from Heidegger's philosophy. At the beginning of his tenure as rector he saw in Hitler an artist capable of ushering in the necessary transformation of philosophical thought and, thereby, the entirety of Western culture.

One of the most troubling affairs associated with his role as rector concerns his treatment of his teacher, Edmund Husserl. In late April 1933, Heidegger signed the letter dismissing Husserl from his position at the university. Although Heidegger did not actually fire Husserl himself and was merely carrying out the decision to dismiss all Jewish faculty members, he made no effort to intervene either, as he had done in other cases. Despite rumors to the contrary, Heidegger was probably not responsible for the decision to bar Husserl from the library, but, once again, he made no effort to have it revoked.

Heidegger's wife, Elfriede, delivered the letter to Husserl, together with a bunch of flowers. The two men never spoke again, and Heidegger did not even attend Husserl's funeral in 1938. Husserl, who was deeply hurt by Heidegger's action and by the fact that he never came to see him after the letter, died alone. It is interesting, in this context, to note the remarks of Hannah Arendt, who loved and admired Heidegger, to Karl Jaspers: "He knew very well that that letter would have left Husserl more or less indifferent if someone else had signed it … And because I know that *this letter and this signature almost killed him*, I can't but regard Heidegger as a potential murderer" (Arendt & Jaspers, 1992, pp. 47–48).[3]

In 1941, apparently under pressure from his publisher, Heidegger removed the dedication to Husserl from *Being and Time*. It is also interesting to note a remark that Heidegger made in 1939, in his *Black Notebooks*, recognizing the lasting importance of the phenomenological turning-point (in relation to psychologism), for which Husserl was responsible, while criticizing Husserl's inability to arrive at essences—probably

[3] Unlike Arendt, who praised Jaspers for his "spiritual" resistance to Nazism, historian Golo Mann—son of the acclaimed German writer Thomas Mann and a student of Jaspers at Heidelberg—noted that Jaspers in fact remained completely silent during the war. So profound and internalized was his opposition to Nazism, Mann sarcastically remarked, that no one even knew about it. Needless to say, Mann cut off all ties with his former teacher.

due to his "race".[4] This remark is entirely in keeping with a well-established Heideggerian position: "The more originary and inceptual the future decisions and questions become, all the more inaccessible will they remain to this 'race'" (Heidegger, 2017, p. 37).

In April 1934, Heidegger submitted his resignation from the rectorate. He continued to pay dues to the Nazi Party (it is worth noting that very few intellectuals who were party members stopped paying their dues), but no longer took part in party activities. On 30 June 1934, following the Night of the Long Knives (Hitler's purge of the SA leadership), Heidegger expressed disappointment with the Nazis and seems to have begun developing the idea of some lost essence of National Socialism that no one else—not even the Nazis themselves—seemed to see.

In 1934–1935, Heidegger's classes, particularly his seminar on Nietzsche, were placed under supervision, and his chances of a public career in the Third Reich began to wane. In the summer of 1944, the university authorities declared Heidegger to be "the professor whose services can be most readily dispensed with" (Steiner, 1987, p. 259).

To summarize, judging by Heidegger's actions and words in the years 1933–1945 (and thus way beyond), there can be no doubt that he did indeed support the Nazi Party and even showed signs of admiring Hitler himself. Heidegger's appearance and writings at the time about the connection he shared with Hitler (they were born in the same year) often appear ridiculous, but they clearly attest to Heidegger's world view. Heidegger saw himself as a philosophical guide to Hitler and the Nazi Party. He not only supported the Nazi Party, but identified an affinity between his philosophy, his early philosophy that is, and Nazi ideology. As we will see below, the *Black Notebooks* shed further light on the connection between Heidegger and Nazism (Fig. 6.1).

[4] Husserl understood the deep problems with psychologism, particularly when it came to the study of consciousness. Husserl was the first to grasp that consciousness cannot be explained by the natural sciences or (psychophysical) psychology, and that a new science must be developed—one that would study the structure of consciousness not as an object in the world (like the methodology employed in the natural sciences), but as a unique subject. The defining characteristic of our consciousness is its intentional structure. We are always thinking about something, grasping something, envying something, focused on something. Consciousness is the structure that allows us to adjust to the world around us. It is what allows us to grasp the world as we do but, as such, it is precisely not an object to be grasped. The method employed in the natural sciences, eagerly adopted in psychophysical psychology, also treats consciousness as an object; but in so doing, it misses the essence of human consciousness. In order to study consciousness, we must stop treating it as if it were an object in the natural sciences. This is the phenomenological turning point that Husserl proposed.

Fig. 6.1 Heidegger and Nazism
Looking back, over 70 years later, we can say that Heidegger was indeed the true shepherd (*den Führer führen*) of the Nazi movement—someone who never turned his back on it, never apologized for it, never doubted its greatness, and even provides philosophical grounds for exonerating the Nazis for the genocide of the Jewish people

The Black Notebooks

Heidegger's *Black Notebooks*, written from 1931 to a year before his death in 1975, were first published in 2014. Some have called the *Black Notebooks* (*Schwarze Hefte*)—a set of 34 small notebooks in which Heidegger wrote down his observations—a philosophical diary. Lapidot (2020) divides views regarding the content of the *Notebooks* into three categories: (1) those that argue that a clear distinction must be made between Heidegger's philosophical writings and his personal musings in his diaries; (2) those that characterize the *Black Notebooks* as bad philosophy (Nazi propaganda, a "falling back" into metaphysics) that should not be considered in the same light as Heidegger's great philosophy; (3) those that consider the *Black Notebooks* crucial to understanding all of Heidegger's philosophy.

The *Black Notebooks* reveal that Heidegger had great hopes for the Nazi movement beginning in 1930. He saw in Nazism a chance to transform all of Western culture—a transformation of the kind he describes in his "deep history" (*Geschichte*), that is, a tectonic historical shift. Heidegger believed that this movement could lead humanity toward "the other beginning" of

Western culture, transforming the very essence of the thinking that gave rise to European nihilism. In the Nazi ascent to power in 1933 Heidegger saw an opportunity for a veritable philosophical revolution, a chance for "great politics", of the kind that changes history.

Heidegger considered the Nazi revolution a philosophical one that would bring back the splendor of the pre-Socratic world—a reference to the philosophy he developed following his own return to the pre-Socratic philosophers, active in Greece before and sometimes in parallel to Socrates (470–399 BCE). The pre-Socratics explored questions pertaining to cosmology and other natural phenomena (the structure of the world, its physical nature, and so forth—earning them the title "inquirers into nature" in the writings of Aristotle), which they sought to explain in terms of universal laws rather than the will of the gods.

In his philosophy, Heidegger sought to return to Heraclitus (late sixth–early fifth century BCE), in particular. Heraclitus is known for his obscure style and is sometimes referred to as "the Riddler". In his fragments, Heraclitus stresses the unity of opposites ("The road down and the road up is one and the same") and change as a central component of reality—one of his best known assertions is that "it is not possible to step twice into the same river" (the famous reply is that it is impossible to step into the same river even once). Heidegger tried to build a bridge between the beginning of "real" philosophy—that is, not the philosophy of Socrates and Plato but that of the pre-Socratics—his own (he did not suffer from excessive modesty and saw himself as the father of a philosophical revolution), and the essence of the Nazi revolution. In order to understand Heidegger's motivation in all of this, it is worth returning to his "deep history"—history that is "never past but stands ever before us" (Heidegger, 1998, p. 240). The philosophy of Socrates and Plato is the beginning of Western culture and, therefore, "we are still living out the consequences of the fundamental modes of thought established there" (Clark, 2011, p. 30).[5]

Heidegger argued that Being, a central concept in his work, entails a fundamental plan, that is, the binding together of "blood and soil" (*Blut und Boden*)—an undoubtedly Nazi term. He also linked the *Volk* (the German people), a critical concept in Nazi ideology ("*Ein Volk, ein Reich,*

[5] As I wrote in the introduction, this is precisely the methodological approach I adopted in Part I.

ein Führer"), to his own fundamental principle of Dasein (being-there).[6] If we think of the concept *Volk* in terms of a German people that sees itself as superior and imbued with a historical mission, there can be no doubt this is exactly how Heidegger saw the German people—the closest people to the ancient Greeks, who lived in the world itself and were closer to authentic being. The Greeks were, to Heidegger's mind, the epitome of a people that dared to begin *ex nihilo* and that is what he thought of the German people in general, and Hitler in particular: as the leader who could begin anew and thus redefine the end.

It is no coincidence that Heidegger returned to Heraclitus. As we have seen, according to Heidegger, the crisis of the West was a result of the metaphysical tradition, from Plato's cave allegory to Nietzsche's will to power. From a Heideggerian perspective, Nietzsche's superhuman does not offer a solution to the nihilistic crisis but is an example of pure will to power and, as such, is the perfect product of the technological mode of being.

Heidegger was thus of the opinion that founding a new philosophy in the present requires a new beginning—one that would not lead to nihilism, the culmination of which is technological being. He saw the possibility of a new beginning for Western philosophy in Heraclitus, and tried to forge a new kind of bond between 1933-Germany and pre-Socratic

[6] The following are a number of examples of assertions by Heidegger on the subject of German Volk, quoted in Lang's *Heidegger's Silence* (1996, pp. 41–42): "Only from the Germans can world historical mediation come—provided that they find and defend what is German" (*Heraklit,* in *Gesamtausgabe,*vol. 55 [summer semester 1943]); "The peril of world... darkening... [will] be forestalled [only] if our nation in the center of the Western world is to take on its historical mission" (*Introduction to Metaphysics*); "We are caught in a pincers. Situated in the middle, our Volk experiences the severest pressure. It is the Volk with the most neighbors and hence the most endangered—and with all this, the metaphysical Volk. We are certain of this mission. But the Volk will only be able to realize that destiny if *within itself* it creates a resonance … and takes a creative view of its heritage. All this implies that this Volk, as a historical Volk, must move itself and thereby the history of the West beyond the center of their future 'happening' and *into the primordial realm of the powers of Being*" (*Einführung in die Metaphysik*, in *Gesamtausgabe*, vol. 40 [summer semester 1935]); "Reflection on the Volk is an essential stage … An uppermost rung of Being will be attained if a 'Völkisch principle,' as something determinative, is mastered and brought into play for historical Da-sein" (*Beiträge zur Philosophie*, in *Gesamtausgabe*, vol. 65 [1936–1938)]); "The 'Fatherland' is Being itself, which from the ground up carries and ordains the history of a Volk as one that exists: the historicity of its history" (*Hölderlin's Hymnen "Germanien" und "Der Rhein"*, in *Gesamtausgabe*, vol. 39 [winter semester 1934/35]).

Greece.[7] The Nazi Party was supposed to have been the agent of this philosophy for the refoundation of relations between global technology and modern humanity—which is also the reason that Heidegger himself sought to link his philosophy to the ideology of the Nazi movement. The refoundation of relations between modernity and technology would allow humans and humanity to be redefined, perhaps saving the West from descending into nihilism.

In this context, it is worth noting that Heidegger's commitment to Nazism was hardly incidental to his thought, which is why it is pointless to seek justifications or to distinguish between the man and his philosophy. As French philosopher and sociologist Jean-François Lyotard argued, in *Heidegger and "the Jews"*, although Heidegger's philosophy must not be reduced to Nazism or even to the Nazism of Heideggerian philosophy, no distinction should be made between the "philosophical" Heidegger and the "political" Heidegger. The two facets of the man are inexorably linked to one another—a link that is already apparent in *Being and Time*: "The greatest thought can lend itself, as such, to the greatest horror" (Lyotard, 1990, p. 57).[8]

Metaphysical Anti-Semitism

Since publication of the *Black Notebooks*, many scholars have linked Heidegger's philosophy not only to Nazism, but to anti-Semitism as well. Peter Trawny (2016), founder of the Martin-Heidegger-Institut at the University of Wuppertal, believes that "Heidegger didn't just pick up these antisemitic ideas, he processed them philosophically" (quoted in Oltermann, 2014).

Heidegger's notebooks include relatively few classically antisemitic statements. Of a total of 1200 pages, fewer than ten refer to Jews, and in the notebooks for the period 1938–1941, the words *Jude*, *Judentum*, and *Jüdisch* appear exactly four times. Based on these diaries, Heidegger would appear to have seen the Jews as a metaphysical threat and the epitome of

[7] This is my interpretation—one with which some Heidegger scholars would certainly disagree (although it probably depends on the years in question).

[8] In this context, the following remark is also interesting: "It is difficult to attribute an apolitical quality to a work like *Sein und Zeit*, of which the entire second section is devoted to the power that *Dasein*, and notably that destiny called *Volk*, has to escape from inauthenticity and to open itself to the future-as-coming-toward of its fate by giving (delivering) to itself the knowledge of its 'having-been'—what is called *historicality* (Lyotard, 1990, p. 67).

nihilism: "the kind of human existence which in an utterly unrestrained way can undertake as a world-historical 'task' the uprooting of all beings from being" (Heidegger, 2017, p. 191). Heidegger even believed that the Jews themselves were responsible for the development of the idea of race: "With their emphatically calculative giftedness, the Jews have for the longest time been 'living' in accord with the principle of race, which is why they are also offering the most vehement resistance to its unrestricted application" (ibid., p. 44). Heidegger's description of this threat—including the claim that the Allies were the carrying out the aspirations of "world-Judaism" (*Weltjudentum*)—is filled with Nazi jargon, combined with the jargon he developed in *Being and Time*.

Many seek to reject the notion that Heidegger was an antisemite in the simple, vulgar sense, contending rather that his anti-Semitism was of a more complex and sophisticated variety. However, even these scholars would find it hard to deny that Heidegger did adopt a kind of metaphysical anti-Semitism, that is, Jew-hatred based primarily on arguments rooted in an overarching metaphysical approach. Indeed, Heidegger explicitly argues that, "the question of the role of world-Judaism is not a racial question, but a metaphysical one" (ibid., p. 191). Before moving on, I will try to present this issue in a richer and more complex fashion.

Within the framework of this metaphysical racism, Heidegger distinguishes between the Germans, with their "spirituality", and the Americans and Russians, who are addicted to "intelligence", photography and representation, that is, to Jewish-technological metaphysics, in which thought is replaced by calculation. This distinction makes the Germans more human, that is to say more worthy of living, while the Russians and the Americans are more "animal", that is to say expendable. When Heidegger speaks of the human race (*Geschlecht*), he refers to the corrupt race, the race that has lost its home, the disoriented race, the deteriorated race, the disintegrated race. In "Language in the Poem: A Discussion on George Trakl's Poetic Work", published in 1952, Heidegger does not identify this concept with any particular people. In his *Introduction to Metaphysics*, however—published in 1935 and intended to serve as the "philosophical foundation of Nazism", to allow Heidegger to become the philosophical guide of Nazism, to "lead the leader"—he identifies the human race that has lost its way with the German people. Heidegger speaks of "our ownmost Dasein, in the whole of the history that is allotted to us" (Heidegger, 2000, p. 44) and of "our Dasein … summoned to its history … called to make a decision in it" (ibid., p. 47). The summons to return home, to find the Dasein allotted to the Germans alone—described by Heidegger as "the people richest in neighbors and hence the most endangered people, and for all that, the metaphysical people". The fate of Europe and the West depends on the fate of the German people, which lies "in the pincers" and must, therefore, "transpose itself—and with it the history of the West". Heidegger was well aware of the apocalyptic possibilities contained within this call to return to "the originary realm of the powers of Being", which he himself qualified—"Precisely if the great decision regarding Europe is not to go down the path of annihilation" (ibid., p. 41). Heidegger thought of Germany in terms of the center and of the pincers as being on the outside. He viewed Russia and America as having fallen deep into the metaphysics of technology: "This Europe, in its unholy blindness always on the point of cutting its own throat, lies today in the great pincers between Russia on the one side and America on the other." Here Heidegger makes sure to emphasize that "Russia and America, seen metaphysically, are both the same: the same hopeless frenzy of unchained technology" (ibid., p. 40). Thus, while the Germans are at the center, closest to Being, to humanity, the Russians and the Americans are far outside, distant from Being and from humanity—unlike Odysseus, unlike the Germans, no longer able to return home. When Heidegger speaks of the Russians and the Americans as having exchanged quality for quantity, spirituality for empty cleverness, and, most of all, spirit for intelligence, he is speaking of those who are outside and can no longer be part of the human race, of those who are like animals that cannot be outside themselves, of those who have lost their world. From Heidegger's perspective, the Americans and the Russians, operating perhaps within the framework of Jewish-technological-calculatory metaphysics, are to blame "for the darkening of the world, the flight of the gods, the destruction of the earth, the reduction of human beings to a mass, the hatred and mistrust of everything creative and free" (ibid., p. 40). What the Americans and the Russians are missing, according to Heidegger is the world itself. Like animals, they are bereft in the world, since, like animals, they have lost the possibility of returning home, to Being that is not dominated by technology. Their path to Being is blocked. Not only have they darkened the world, but they themselves are spirit-less and world-less. And "As soon as they are expelled into the animalistic outside," explains Dror Pimentel, "they are fair game." Accordingly, "In relegating Russia and America to animality, Heidegger is giving their eradication his philosophical endorsement" (Pimentel, 2019, p. 247). The superiority of the German people, according to Heidegger, is a matter of spirituality rather than race ("The German people is more elevated than any other because it has appropriated spirit"), although the end result is the same—according primacy to the German people and sanctioning the eradication of peoples far from the center that have lost the ability to return home. Heidegger may thus be said to have embraced an overarching metaphysical approach according to which it is the Germans, as the most spiritual of all peoples, who have the ability to deliver the West from nihilism, that is, from falling into technological being; and it is the Jews, as the inventors and embodiment of technological-calculative thinking, who have made the world alien and other. This is exactly what Heidegger means when he refers to "world-Judaism": a landless people that feels alienated in the world. Heidegger describes the end of the West as a situation in which Judaism will determine the global agenda—a situation in which it will no longer be possible to feel at-home in the world.

To return to Heidegger's metaphysical anti-Semitism, the rhetorical question posed by one of his foremost interpreters, Jewish-Algerian-French philosopher Jacques Derrida, now becomes clear: "Is a metaphysics of race more or less serious than a naturalism or a biologism of race?" (Derrida, 1989, p. 74). Indeed, some have suggested that the metaphysical anti-Semitism (traces of which can also be found in the writings of Kant and Hegel)[9] that Heidegger developed is no less serious than "ordinary" anti-Semitism, because ultimately, for Heidegger, overcoming "Jewish" metaphysics required the elimination of the Jews (Di Cesare, 2018). Indeed, various assertions in the *Black Notebooks* and elsewhere show that there is a real possibility that Heidegger was also an "ordinary" antisemite. For example: "World-Judaism … cannot be held fast anywhere and, with all its developed power, does not need to participate anywhere in the activities of war, whereas all that remains to us is the sacrifice of the best blood of the best of our own people" (Heidegger, 2017, p. 208). Kravitz argues that "Heidegger's assertions in this vein, accusing Judaism of 'rootlessness', 'landlessness', or 'empty rationality', simply echo old, familiar stereotypes. In this sense, Heidegger was a rather ordinary antisemite—common, banal, and wholly devoid of 'authenticity'" (Kravitz, 2019, p. 27). In his book *The banality of heidegger* the French philosopher Jean-Luc Nancy goes a step further, arguing that Heidegger's assertions in the *Notebooks* are based on *The Protocols of the Elders of Zion*. According to Nancy, it is inconceivable that Heidegger, as an academic, would not have

[9] For an extensive discussion, see (Rose, 1990).

known that the book was an antisemitic forgery—making his anti-Semitism all the more crass (Nancy, 2017).

Others argue that Heidegger's anti-Semitism was even more radical. If the Jews symbolize the collapse of thought and its replacement with calculation, then, from a Heideggerian perspective, they stand at the heart of the fall of the West. Heidegger's anti-Semitism might thus be said to be even more systematic than that of the Nazi regime. Lapidot (2020) develops this idea and suggests that there is a profound connection between Heidegger's thought concerning Jews and his thought concerning technology—that is, as a fundamentally nihilistic super-metaphysics that precludes the possibility of being in the world. In order to fully understand this argument, we must first understand the essence of technology from a Heideggerian perspective.

The Essence of Technology According to Heidegger

The essence of technology, according to Heidegger, "is by no means anything technological" (Heidegger, 1977, p. 4). It is not a thing, or a tool, or something that can be grasped, but a total system, an order of things (as well as practices and techniques) into which we are thrown and within which we act and are acted upon. When Heidegger speaks of modern technology, he is, in effect, speaking of the modern human condition. It is not that technology is a characteristic of the modern condition, but that technology and the modern condition are (virtually) synonymous.

Heidegger argued that the technological threat does not stem from technology itself: "What is dangerous is not technology. There is no demonry of technology" (ibid., p. 28). The danger lies not in the destruction of nature and culture as a result of our use of various technologies, but in transparency and forgetting—both inherent to the essence technology. Technology prevents us from understanding how we were thrown into the world. Humans are thrown into the world through the complete transparency of technology, without understanding that technology not only mediates and explains the world but also creates it (in that sense, to say that technology is a deterministic system limits its real power; it is more like a field of gravity). The danger is thus the danger of forgetting. In the context of our complete absorption in technological being, humans lose the ability to reveal their original thrownness-to-the-word. In short, technology changes everything, including humans themselves (and, no less importantly, the very concept of humanity), and does so in a concealed manner, like water for fish.

Within the framework of the technological mode of being, we have lost the ability to see nature as it is. We walk along marked paths, clothed in technology that protects us from the elements, through defined and delimited areas, observing animals bred in captivity—which we then seek to return to nature which has already become technology. Nature no longer exists, only nature reserves—thus it would be more accurate to say that because we have lost the ability to see nature as it is, it became a resource and nothing more.

Technology challenges and redefines nature: "Air is now set upon to yield nitrogen, the earth to yield ore, ore to yield uranium, for example; uranium is set upon to yield atomic energy, which can be released either for destruction or for peaceful use" (ibid., p. 15). Heidegger further stresses that "the hydroelectric plant is not built into the Rhine River as was the old wooden bridge that joined bank with bank for hundreds of years. Rather the river is dammed up into the power plant. What the river is now, namely, a water power supplier, derives from out of the essence of the power station." The river remains a river in the landscape, "in no other way than as an object on call for inspection by a tour group ordered there by the vacation industry" (ibid., p. 16). From a technological perspective, the world is nothing but a giant petrol station, a resource for the creation of energy that is produced, stored, transported, and made available for use.

The challenge of nature, or the "challenging claim" (ibid., p. 19) is the essence of modern technology (*das Ge-stell*, Enframing).[10] It is a process that turns everything into a resource—at its end point, even humans become "things": technology enframes humans in the world, transforming them into a "standing-reserve" (*Bestand*), raw material: Modern technology as an ordering revealing is, then, no merely human doing. Therefore we must take that challenging that sets upon man to order the real as standing-reserve in accordance with the way in which it shows itself. That challenging gathers man into ordering. This gathering concentrates man upon ordering the real as standing-reserve. The "thing", the piece, is not

[10] In this, Heidegger was influenced by Jünger, who spoke about the *Gestalt* of the worker in terms of total mobilization (*Die totale Mobilmachung*): "By *Gestalt* we refer to the highest meaningful reality. Its appearances are meaningful as symbols, representations and impressions of this reality. The *Gestalt* is the whole which embraces more than the sum of its parts. This 'more' we call totality" (translated in Zimmerman, 1990, p. 57).

part of a whole, but stands by itself. Thus, within the framework of technology (technological being), humans are thrown into the world without significance or value, no more than "things" and, as such, can never be missing, because every part is replaceable; that is its essence:

> The consumption of all materials, including the raw material "man," for the unconditioned possibility of the production of everything is determined in a concealed way by the complete emptiness ... Technology has to ... create a substitute, and consume the raw materials. But in truth the "substitute" and the mass production of ersatz things is not a temporary device ... This circularity of consumption for the sake of consumption is the sole procedure which distinctively characterizes the history of a world which has become an unworld. (Heidegger, 1973, pp. 106–107)

The framework within which technology places nature and humanity does not allow us to think in terms other than those of maximum exploitation, although humans, who play a *dual role*[11] in relation to technology—acting not merely as a standing-reserve but also developing and employing it—will never be "mere stock" (Heidegger, 1977, p. 17). Within the context of technological being, however, even humans exist as a "standing-reserve" (even if they are those who changeless nature, they are in fact challenged themselves to do so), and it is here that the danger described by Heidegger presents itself: as soon as man sees himself "exclusively as standing-reserve ... then he comes to the very brink of a precipitous fall" (ibid., pp. 26–27). It is important to stress, at this point, that according to Heidegger, "the saving power does most profoundly take root and thence thrive even in that wherein the extreme danger lies" (ibid., p. 29). In other words, it is only where we stand as a reserve, that is, in the depths of technological being, that salvation—a free relationship to technology—can emerge:

> Where the danger is as the danger, there the saving power is already thriving also. The latter does not appear incidentally. The saving power is not secondary to the danger. The selfsame danger is, when it is as the danger, the saving power. (ibid., p. 42)

[11] "Who accomplishes the challenging setting-upon through which what we call the real is revealed as standing-reserve? Obviously, man... Only to the extent that man for his part is already challenged to exploit the energies of nature can this ordering revealing happen. If man is challenged, ordered, to do this, then does not man himself belong even more originally than nature within the standing-reserve?" (Heidegger, 1977, p. 18).

Heidegger thus identifies a deep and inherent connection between the danger associated with technological being and salvation (release from this mode of being)—a position he appears to derive not from any historical study or philosophical approach, but solely from the poetry of Friedrich Hölderlin (1770–1843), widely considered one of the greatest German poets.

In any event, those who open themselves to the essence of technology also reveal their own essences, in the sense of revealing truth, since "Being [is] the essence of technology" (ibid., p. 38). It is precisely through this process of recognizing the essence of technology that we are able to establish a free relationship to it, to release ourselves from its grip.[12] As we shall see, Heidegger himself wavered between the view that humans can act to establish a free relationship to technology and (in the later stages of his life) the view that there is nothing we can do but wait until technology destroys itself.

Let us try to understand the possibility that humans may establish a free relationship to technology—perhaps the most complex and explosive question in Heidegger's later philosophy. When humans reveal themselves as standing-reserves, ready to be used—that is, when they reveal their mode of thinking and the essence of current historical thought—that is precisely the point at which they reveal the essence of their own Being and embark on the path to release from technology. The thinking of one who embarks on the path to emancipation is no longer subject to the logic of challenging alone, but is open to other possible ways of thinking, like those found in poetry and art.

Here I would ask whether the Heideggerian philosophy of technology enables the exoneration of the Nazi movement from its responsibility for the genocide of the Jewish people. In order to answer this question, I will turn to Heidegger's famous remark from December 1949 (Bremen lecture):

> Agriculture is now a motorized food industry, the same thing in its essence as the production of corpses in the gas chambers and the extermination camps, the same thing as blockades and the reduction of countries to famine, the same thing as the manufacture of hydrogen bombs. (Heidegger, 2012, p. 27)

[12] In Heidegger's later philosophy, the central concept used to describe the process of breaking free from technology is *Gelassenheit* (releasement), which entails the search for an alternative to the mode of thinking that characterizes modern technological being, as well as emancipation from the technological will that Heidegger terms the "will to will".

This remark would seem to indicate, as I will show at greater length in the next subsection, that Heidegger's ideas about technology were never detached from his disappointment in Nazism and his political involvement. To put it more bluntly, Heidegger's ultimate goal (in the sense of telos) in *The Question Concerning Technology* is to absolve the Germans of national and personal responsibility for the Holocaust and, as we shall see, to blame the Jews.

"Agriculture Is Now a Motorized Food Industry, the Same Thing in Its Essence As the Production of Corpses in the Gas Chambers and the Extermination Camps"

Heidegger tried to explain his behavior in a number of documents: (1) a report he sent to the Denazification Committee on 4 November 1945[13]; (2) a 40-page text he wrote in 1945 and entrusted to his son Hermann—published in 1983; (3) an interview he gave to the German weekly *Der Spiegel* in 1966, published, on Heidegger's request, only after his death in 1976 (see Fig. 6.2). The following are his arguments—most of which, I would add, have been refuted by historical scholarship.

(a) His support for the Nazi Party stemmed from a desire to defend the country at a time of general chaos, including the communist threat.
(b) He did not want the rectorship, and only accepted the role in order to safeguard the stability of the university as an independent institution. Heidegger even claimed that he had accepted the position in order to prevent it from being taken over by a Nazi functionary. Scholars have found absolutely no evidence of this, and the claim would appear to be without basis.
(c) His rectorship address was misinterpreted—although, as we have seen, it does include numerous expressions clearly associated with Nazi ideology. Heidegger's claim to *Der Spiegel* that his use of the word *Kampf* referred to the metaphysical struggle described by Heraclitus is ridiculous. Even if Heidegger himself saw the connection to Heraclitus, and even if such a connection did indeed exist, it is unlikely that many in the audience would have been familiar

[13] On the recommendation of Karl Jaspers (whom Heidegger gave the committee as a reference), the Denazification Committee decided that Heidegger would be barred from teaching until the 1950/51 academic year but would continue to receive a monthly stipend.

Fig. 6.2 Heidegger's *Der Spiegel* Interview (1966)
In his interview to *Der Spiegel*, Heidegger was asked: "In 1935, in a course that in 1953 was published as *Introduction to Metaphysics*, you said: 'What today'—this was, therefore, 1935—'is bandied about as the philosophy of National Socialism but has absolutely nothing to do with *the inner truth and greatness of this movement* (*namely, with the encounter between technicity on the planetary level and modern man*) casts its net in these troubled waters of "values" and "totalities".' Did you add those parenthesized words for the first time in 1953, that is, at the time of the publication, in order to explain to the reader of 1953, so to speak, in what way you saw the 'inner truth and greatness of this movement' (i.e., of National Socialism) in 1935—or did you have this explanatory parenthesis already there in 1935?" Heidegger replied: "The parenthesis stood in my [original] manuscript and corresponded precisely to my conception of technicity at that time, and not yet to the later explication of the essence of technicity as 'pos-ure' (*Ge-Stell*). The reason I did not read the phrase publicly was that I was convinced of the proper understanding of my listeners, although stupid people, informers and spies understood it differently—and also wanted to" (Heidegger, 1981, p. 55)

with the text in question, although most would certainly have been familiar with Hitler's *Mein Kampf*—a fact that Heidegger would have been well aware of.[14] If anything, the address actually deepens the connection between Heidegger's philosophy and the Nazi movement. In any event, we have already seen considerable evidence of Heidegger's willing support for Hitler, beyond any formal obligation.

(d) Heidegger claimed that he resigned in 1934, because he refused to accept the line dictated by the regime or the demand that he dismiss two deans. There is, in fact, evidence to the contrary. First, he was not in conflict with his superiors, and, second, he was never asked to dismiss the deans in question (in one of the two cases, evidence of that is very strong). His resignation appears to have been due to his failure to achieve the kind of power he had hoped for—to become the philosophical shepherd of the Nazi movement.

(e) After his resignation, he was targeted by the Nazi regime, which did not allow him to publish his books. The regime indeed appears to have prevented the reprinting of his books.

(f) He claimed that he had always disapproved of biological racism. As we have seen, however, this in no way demonstrates that he was not a metaphysical antisemite.

Heidegger's responses to Nazism are generally divided in two: the philosophical and the personal. The division seems clear: silence on a personal level, and little response on a philosophical level. In this subsection, I will show that both Heidegger's silence and his philosophical responses attest to an unequivocal position, constituting a single vector.[15]

Some of Heidegger's post-1945 remarks pose considerable difficulties. In his first lectures after his suspension, delivered in Bremen (not in a university setting) in 1949, Heidegger related to Nazism in the following manner:

[14] As early as 1931, Heidegger sent his brother a copy of *Mein Kampf* for Christmas, praising Hitler's "extraordinary and unwavering political instincts. See Adam Soboczynski and Alexander Cammann, "Heidegger and Anti-Semitism Yet Again", *LARB*, 25 December 2016.

[15] I am generally averse to the psychologization of science and philosophy—such as the idea that Darwin's theory of evolution was influenced by his relationship with his father. It seems ridiculous and trite. In Heidegger's case, however, considering his philosophical approach to silence, it is hard to separate his personal silence from his philosophical writing.

1. "Agriculture is now a motorized food industry, the same thing in its essence as the production of corpses in the gas chambers and the extermination camps, the same thing as blockades and the reduction of countries to famine, the same thing as the manufacture of hydrogen bombs." (Heidegger, 2012, p. 27)
2. "Hundreds of thousands die in masses. Do they die? … They become pieces of inventory of a standing reserve for the fabrication of corpses. Do they die? They are unobtrusively liquidated in annihilation camps. And even apart from such as these—millions now in China abjectly end in starvation." (ibid., p. 53)

From the first remark, we can deduce that from an essential perspective ("identical essence"), Heidegger:

(a) Saw no difference between the actual murder of people and the creation of technology capable of destroying people but without having done so in practice (hydrogen bomb).
(b) Made no distinction between the murder of people in gas chambers (he does not mention the word "Jews", and not by chance) and mechanized agriculture.
(c) Made no essential distinction between wartime siege and the planned persecution of a particular race, the Jewish race, with the goal of annihilating it.

The reason that Heidegger saw no essential difference between these things is that he believed that all of the cases cited above were the result of living within the framework of total technological being—that is, within Enframing (*Ge-stell*).

It is also worth noting what Heidegger did *not* say:

(a) That from a Nazi perspective, the processes may be similar, but from a moral standpoint, there are clear differences between them.
(b) That although the murder of Jews in the gas chambers and the subsequent disposal of their bodies in the crematoria were conducted in an industrial fashion, he found the actions themselves deeply disturbing, since they were, in many ways, the greatest of the horrors perpetrated by the Nazis.
(c) That the Nazis murdered two million Jews in far less sophisticated ways.

(d) That the victims of these actions were Jews—something he fails to mention in either of the above quotes: "the same thing in its essence as the production of corpses [*of Jews*] in the gas chambers"; "They [*the Jews*] are unobtrusively liquidated in annihilation camps".

The reason that Heidegger did not say these things is simple. From his perspective, all of these things (agriculture, concentration camps, and hydrogen bombs) are part of the fall into the metaphysics of technology, that is, calculative thinking, or, the Jewish mode of being. Consequently, not only are the Nazis exonerated from the murder of the Jews, but the mode of being that brought it about—technological being—is, in fact, a Jewish "creation". That is to say, the Jews were the authors of their own deaths, a well-known Nazi claim.

Even today, after all we know about Heidegger, there are those who insist on presenting his position (based on the first part of the remark: "Agriculture is now a motorized food industry, the same thing in its essence as the production of corpses in the gas chambers and the extermination camps") as a radically ethical and non-anthropocentric stance that views all murder as a tragedy. This attempt does not withstand the test of reality, however. Heidegger simply did not think in such terms and, unlike his Jewish student Hans Jonas, was not disturbed in the slightest by the meat industry and ecological questions (Jonas, 1984).[16] In any event, when this remark is approached in the context of other such expressions, it is clear that Heidegger was not disturbed by the genocide of the Jewish people, which was, from his point of view, the realization of technological metaphysics, the consequence of calculative thinking—Jewish thinking, to his mind, perhaps even Jewish essence. In short, Heidegger blamed "Judaism" for the fall of the West into technological being—the ultimate outcome of calculative thinking. The essence of Judaism and the essence of technology were, from a Heideggerian perspective, one and the same. The essence of technology, which is, in fact, the essence of the modern era, is the principle of total efficiency, within the framework of

[16] Neumann (2019) stresses that there is no contradiction between the Nazi ecological approach and the concentration camps. On the contrary, Auschwitz was the ultimate realization of Nazi ecologism, which viewed the Jews as a pollutant of the pure and pristine fatherland and thus an inherently ecological problem. The Jew is "a parasite that lives at the expense of nature ... unnaturally and anti-naturally" and, as such, must be annihilated, disposed of, as one would dispose of inorganic waste. Hence the perception of Auschwitz as a human waste site, the "*anus mundi*".

instrumental rationalism. Thus, from a technological point of view, it matters little whether we are discussing the egg industry, fast-food hamburgers, satnavs, or Jews sent to the gas chambers and the crematoria. In all of these cases we see the aspiration to the complete exploitation of resources (and their reuse), based on the principle of total efficiency—the ultimate consequence of calculating, dealing, Jewish thinking. Heidegger's argument can also be expressed as follows: the moment the Jewish problem arose, the techno-engineering solution that presented itself was the virtually inevitable consequence of the technological mood, that is, let us stress once again, the "Jewish" mood. The Jews, in the context of Heidegger's overarching metaphysical approach, are responsible for the collapse of the spirit in favor of calculation, thereby leading to the nihilistic crisis and to their own annihilation by the method that they themselves had created. It is worth noting here that Heidegger also attributed the invention of the principle of race to the Jews: "With their emphatically calculative giftedness, the Jews have for the longest time been 'living' in accord with the principle of race, which is why they are also offering the most vehement resistance to its unrestricted application" (Heidegger, 2017, p. 44). Together, these two insights form a single argument that leaves no real doubt as to who is responsible for the genocide of the Jewish people: it is the Jews themselves. This point will be discussed at greater length below.

The second remark ("Hundreds of thousands die in masses. Do they die? … They become pieces of inventory of a standing reserve for the fabrication of corpses. Do they die? They are unobtrusively liquidated in annihilation camps. And even apart from such as these—millions now in China abjectly end in starvation") can be read as a profound observation regarding the danger inherent to technology: the idea that mass industrialized death has become a trivial matter and that the concentration and annihilation camps brought the transformation of human beings into a "standing-reserve" (a resource destined for exploitation and final utilization) to its end point.

According to Heidegger, technological being or, more precisely, technological thinking (thinking as calculation), reached its culmination in the concentration and annihilation camps, where humans themselves were transformed into mere raw material. Italian jurist and political philosopher Giorgio Agamben echoed this idea, observing that "in Auschwitz, people did not die[17]; rather, corpses were produced ... And, according to a possible and widespread interpretation, precisely this degradation of death constitutes the specific offense of Auschwitz, the proper name of its horror ... Death at this point is a simple epiphenomenon" (Agamben, 2002, pp. 72, 86). Yet, while Agamben sought the voice of the survivor, out of the fear that humanity itself, as a concept and a physical reality, had collapsed in Auschwitz, Heidegger was never interested in the victims, let alone in giving them voice. It is also possible that Heidegger's profound insights stemmed from an inability to see the human, the subject, the Jewish victim. It is worth noting that Heidegger was not consistent and had no problem whatsoever seeing the German subject. The question thus arises whether he may not simply have believed that Jews are not human, categorically speaking.

If the two remarks are considered as a single unit, not a word is said in either about the perpetrators. When it comes to the genocide of the Jewish people, Heidegger has no interest in those who planned and carried it out. He sees the Holocaust as an expression of European nihilism—and there may be some profound insight in that—but, at the same time, he is clearly evading "specific German responsibility for the murders" (Clark, 2011, p. 130). Even if Nazism was related to European nihilism, members of the German nation conceived, planned, and carried it out, and they did not seek to annihilate all "people", but "Jews".

When Heidegger does relate to the Holocaust, it becomes clear, time after time, that he does not see the Jews as victims. His reply to Marcuse's appeal, on 28 August 1947, that he at least denounce the acts of the Nazis after the fact and apologize for his own involvement speaks for itself:

> To the [grave, justified accusations][18] that you express "about a regime that murdered millions of Jews, that made terror into an everyday phenomenon, and that turned everything that pertains to the ideas of spirit, freedom and truth into its bloody opposite," I can merely add that if instead of "Jews" you had written "East Germans" [i.e., Germans of the eastern territories], then the same holds true for one of the allies, *with the difference that everything that has occurred since 1945 has become public knowledge, while the*

[17] It is worth asking, why "people" and not simply "Jews"? Here too, as in many other places in the book, I am grateful to Amit Kravitz, for his particularly keen eye.

[18] I have changed Wolin's translation here, in keeping with the original German.

bloody terror of the Nazis in point of fact had been kept a secret from the German people.[19] (Heidegger & Marcuse, 2004, p. 266)

This response is sickening. When referring to the murder of Jews, Heidegger limits himself to quoting Marcuse—as if unable to write these things in his own words. We must also not ignore his use of a particularly dishonest argument made by many Germans, as if the genocide had been hidden from them, since we know that there was a concentration camp near the village where he grew up and another near the university where he worked. Worst of all, perhaps, as American historian Richard Wolin suggests, this quote, taken together with various remarks in the *Black Notebooks*, indicates that *Heidegger actually saw the Germans as the true victims of the war*, although he sustained that all of the things that had been concealed between 1939 and 1945 were now known (Wolin, 2015).

Heidegger's way of dealing with the horrors is to adopt a "perspective from nowhere", a perspective from which human life is entirely without value—unless, of course, you are a German, as we have seen. From this perspective, there is no difference between the food industry and the gas chambers—and I would stress that it is impossible to postulate a unique Heideggerian ethic on this point. Heidegger does not compare the phenomena to awaken us to the horrors of the meat industry, but exactly the opposite: he is not disturbed by the gas chambers in which Jews were actually murdered, just as he is not disturbed by chicken coops or cattle farms. This is not silence. It is his explicit position. In any event, were Heidegger consistent and were he to maintain his perspective from nowhere even when it came to Germans, we could at least credit him with trying to present philosophical arguments regarding the essence of technology and humanity. The fact that every time it comes to Germans he is perfectly

[19] This brief letter (Heidegger to Marcuse, 20 January 1948) seems to represent Heidegger's most sincere response on the subject. His silence is not authentic and has no meaning beyond his desire to protect himself and his family. He thus writes in his third point, "You are entirely correct that I failed to provide a public, readily comprehensible counter-declaration; it would have been the end of both me and my family. On this point, Jaspers said: that we remain alive is our guilt" (Heidegger & Marcuse, 2004, p. 265). Even here, there is no real apology. The statement "that we remain alive is our guilt" is too general. It is something that anyone could say, a Dostoevskian utterance that offers no real assumption of responsibility. It is Levinas who proposes an ethical alternative, going back to the original quote from *The Brothers Karamazov*: "We are all guilty toward others and I am the guiltiest of all."

capable of ascribing unique individuality to the subject merely shows just how twisted this way of thinking is, how dangerous a philosophy it was that sought only to exonerate the German people.

At this point we must ask: if Heidegger was not only a pitiful human being, but also a lousy philosopher—at least when it comes to his clumsy attempt to use philosophy to absolve the Germans of responsibility for the genocide of the Jews—why should we read Heidegger at all? The answer is quite simple. There is a real possibility that Heidegger's interpretation of Plato and Aristotle on the matter at hand is not entirely unfounded and is even incisive in many ways, while the connection he makes between these ancient philosophers and Nietzsche is not only possible but necessary. This topic deserves a book of its own,[20] but we can at least try to examine this argument (perhaps in a more moderate form) through the criticism expressed by the later Heidegger regarding the source of the essence of modern technology already in the way in which Plato and Aristotle understood the fundamental structure of philosophy. According to Heidegger, Plato and Aristotle viewed ontology (questions about reality—about what is in the world and about the meaning of "being") as the conception of all phenomena in terms of creative production or poiesies (Gr. ποίησις). They thus perceived the world of phenomena as if it were a world of products. In order to understand technology and the way in which it operates in the current era, claims Heidegger, we must consider the concept of production or "bringing-forth" in its broadest sense, since it lies not only at the root of Greek thinking, but at the root of modern thinking as well, in keeping with the essence of technology: "Through bringing-forth, the growing things of nature as well as whatever is completed through the crafts and the arts come at any given time to their appearance" (Heidegger, 1977, p. 11).

If it is indeed the case that the metaphysics of technology links Plato and Aristotle to Nietzsche, and nihilism derives from the foundations of Western thought, then: (1) We must revisit the writings of all of the important philosophers, on the assumption that we have missed some essential aspect of their thought that Heidegger can help us discover; (2) We must seriously consider the argument that the connection between Western philosophy and the concentration camps is not entirely a

[20] For an extensive discussion, see (Weinberger, 1992) and (Skrbina, 2015). On Heidegger and the history of "productionist" metaphysics see (Zimmerman, 1990, pp. 156–190). See also (Pippin, 2002).

coincidence; (3) It stands to reason that the danger posed by this mode of thinking is not safely in the past, that Heidegger and Nazism were merely an expression of the philosophical thinking that shapes Western culture in a variety of ways to this day.

Many would be willing to forgive Heidegger for his actions in the years 1933–1945 (or, more specifically, in the years 1933–1934) and to dismiss them as political foolishness, if not for the fact that beyond the few remarks we have discussed here, he chose to keep silent. And indeed, his true position is slowly coming to light. His words and silences were equally considered.

Some suggest that Heidegger's silence should be respected, because he understood that the event (the Holocaust in general and his own involvement in particular) was beyond the realm of language, as many scholars of trauma have claimed. I would point out, however, that this argument should be reserved for the victims. As Shoshana Felman and Dori Laub wrote in *Testimony* (1992), those who have experienced extreme trauma may choose silence or fall into silence.[21] That is not the case, however, with perpetrators. Heidegger's silence signifies agreement, and if the Jewish poet Paul Celan—who shared a deep bond with Heidegger (one might even say mutual admiration)—could find the words, then Heidegger should at least have tried (or, in the words of Celan, "We have exchanged a good deal of mutual silence. I think that someday some of it will be redeemed from unspokenness through conversation"; quoted in Lyon, 2006, p. 189), even at the risk of achieving an imperfect result.[22] In a

[21] For further debate, see (Ataria, 2016, 2017, 2018, 2021).

[22] Since we have little reliable information about what occurred at the meetings between Heidegger and Celan in the years 1967–1970, before Celan's suicide, I will not address that issue here. Despite numerous attempts to understand what was said between the two (perhaps it would also be fair to say, attempts to blame Heidegger for Celan's suicide), Steiner stressed that "there is not a shred of evidence that allows us to say that Celan parted with Heidegger defeated and hurt" (quoted in Ruth Almog, "Celan Visited Us. Poor Man, He's not Well" [Heb.], *Haaretz*, 27 February 2002), and it is very hard to link Celan's suicide to his much-discussed meeting with Heidegger. It is difficult to say what Celan meant by the lines he wrote in the visitors' book of Heidegger's "hut" at Todtnauberg, in the Black Forest: "Into the hut-book, with the view of the well-star / with hope for a coming word in the heart." The poem "Todtnauberg" that Celan subsequently wrote (Frankfurt, 1 August 1967) and sent to Heidegger (12 January 1968), echoing the time he spent with the philosopher that day in the Black Forest, sheds no further light on the visitors' book inscription. Nor do we know what was said between the two on the return voyage from Todtnauberg. The theory that Celan, leafing through the visitors' book, discovered the names of ardent Nazis among the hut's previous visitors, is quite possible, although it would be quite a stretch to suggest a link between that and the poet's suicide.

choice between everything and nothing, Heidegger should at least have opted for almost, or, in his own words, "For the most part, we talk too much, but sometimes we also talk too little" (Arendt & Heidegger, 2004, p. 58).

Had Heidegger kept silent after the war, publishing absolutely nothing, his silence might have been worthy of respect. The fact that he wrote and spoke endlessly leads us to the conclusion that his was a thundering silence, a deadly, murderous silence, signifying a refusal to internalize the moral ramifications of the Holocaust. The Jewish philosopher and literary critic George Steiner wrote in this context that, "nauseating as they are, Heidegger's gestures and pronouncements during 1933-34 are tractable. It is his complete silence on Hitlerism and the holocaust after 1945 which is very nearly intolerable", adding that "for a philosopher, for a German witness, for a thinking, feeling human being implicated in at least a part of the relevant events, to say absolutely nothing is tantamount to complicity" (Steiner, 1987, pp. 123–124). The French-Jewish philosopher Emmanuel Levinas wrote of Heidegger's silence: "But doesn't this silence, in time of peace, on the gas chambers and death camps lie beyond the realm of feeble excuses and reveal a soul completely cut off from any sensitivity, in which can be perceived a kind of consent to the horror?" (Levinas, 1989, p. 487).

A thorough examination thus reveals that Heidegger never really recognized the horrors of the Nazi regime and appears to have believed in its positive essence—at least as he thought of it, that is, as the possibility of a renewed encounter between global technology and modern humanity, as a poetic of emancipation—to the end of his life.

In Heidegger's defense, one might argue that, on a personal level, he took no position, certainly not an ethical one. To his mind, the West had capitulated to its own metaphysics (both capitalism and communism were further evidence of this) and, in this sense, industrial agriculture and the concentration camps were merely expressions of this capitulation. There is another possibility, however, much harder to digest: that all of Heidegger's later philosophy, all of his writing on technology, was a product of his political approach (complete identification with Nazism) and his metaphysical anti-Semitism—probably far worse than any "primitive" anti-Semitism (on religious grounds, e.g.,)—and part of an overarching strategy to exonerate the Germans of their responsibility for the genocide of the Jewish people.

At no time was Heidegger disturbed by the genocide of the Jewish people, and he never used the words "Holocaust" or "Jewish genocide". He was disturbed, however, by the fact that the Nazi movement did not live up to its historical mission, as he perceived it—in which, let us recall, he identified danger as well as greatness. For that reason, in only his third reference to Nazism, in *Introduction to Metaphysics*, published in 1953 based on lectures delivered in 1935, Heidegger chose, despite the years that had passed, to leave the following remark:

> In particular, what is peddled about nowadays as the philosophy of National Socialism, but which has not the least to do with the inner truth and greatness of this movement [namely, the encounter between global technology and modem humanity], is fishing in these troubled waters of "values" and "totalities" (Heidegger, 2000, p. 213).

This quote shows that Heidegger always saw in Nazism a movement with a profound philosophy. This is true not only of the years 1934–1935, perhaps swept up in the general enthusiasm of many intellectuals in Germany—but even 8 years after the war, when it was absolutely clear that the Nazi regime had been responsible for the twentieth century's most horrific war and for the organized genocide of millions of innocent people. Thus, even in 1953, when these things were already known, Heidegger continued to believe that the essence of Nazism was the establishment of a new relationship between modern humans and technology—release from subjugation to technology, in favor of existence alongside technology in a state of awareness and freedom. Toward the end of his life, Heidegger believed that emancipation from technology does not depend on the choice of human society to act or to refrain from acting, but on awaiting the arrival of the "last god" and, with it, another beginning for Western culture, that is, the beginning of a different way of thinking and engagement with being itself.[23]

[23] Who is the "last god", according to Heidegger, and what is the "other beginning" we must await? A discussion of these concepts, which are vague and therefore open to interpretation, is beyond the scope of the present discussion (for wider discussion see: Rojcewicz, 2006). I will only note that the idea of another beginning and the return of the last god is first discussed at length in a book central to Heidegger's later thought, *Contributions to Philosophy* (Heidegger, 1999), written between 1936 and 1938. Let it suffice to say here that the last god is necessarily the ultimate expression of the central characteristic of Western thought: understanding individual being through the Supreme Being (a tendency that Heidegger terms "onto-theology"; from the Greek ὄντος, being, and θεός, god). Whether it is an idea, a substance, a monad, or a will to power, it is the Supreme Being that purports to explain all individual cases. This kind of thinking is the legacy of metaphysics and, therefore, of modern technology. It is the basic mechanism, if you will, that leads to understanding humanity and nature in terms of calculation, exploitation, and extraction.

It is worth stressing here that from Heidegger's perspective, emancipation from the domination of technology as a state of being and way of thinking is equivalent to emancipation from Jewish being. It is for this reason that some have claimed that Heidegger's systematic, metaphysical anti-Semitism is no better and perhaps even worse than classic anti-Semitism, from which Heidegger was not altogether free either (in the *Black Notebooks*, for example, he describes Judaism as possessing "the persistent skillfulness for calculating and hustling and shuffling").[24] By that time, however, it was eminently clear that National Socialism had never been interested in establishing a new relationship between humans and technology, but on the contrary, was, from the very beginning, interested in exploiting technology to the utmost, in order to create an exceptionally advanced war machine. If technology, at its most dangerous, is the will to power for the sake of power, then the Nazi regime embodied the total aspiration to power (which, in the framework of nihilism, became the will to will), and Heidegger himself was well aware of this: "The Nazi movement is a symptom for the tragic collision of man and technology, and as such a symptom it has its 'greatness,' because it affects the entirety of the West and threatens to pull it into destruction" (Lapidot, 2020, p. 113). Rather absurdly, from a Heideggerian perspective, one might even say that the Nazi regime, despite its great potential and the hopes that Heidegger had placed in it, ultimately became the most "Jewish" of projects.

If so, even in 1953, when Heidegger already knew without a doubt everything there was to know about the horrors of Nazism, he still chose to remain loyal to what he perceived as the original truth of the Nazi regime. In a certain sense, the Nazis themselves were not sufficiently loyal to that greatness, to that truth.

Here I would like to pose the central question: is what lies, a priori, at the basis of Heidegger's thought about technology his desire to absolve the Nazi movement of responsibility for the horrors of the Second World War and especially the attempt to annihilate the Jewish people? As we shall soon see, Heidegger goes even further, seeking to blame the Jews themselves for their own murder. Of course, everything is argued on a purely metaphysical level, but, as we have seen, metaphysical anti-Semitism is no less radical than "primitive" anti-Semitism.

[24] Quoted in (Lapidot, 2020, p. 116).

Exonerating the Nazis from the Genocide of the Jewish People

In the 1960s and 1970s, Heidegger did not deny that he saw in the rise of Nazism the potential for profound change in the course of Western history. Moreover, he appears to have seen his own philosophy as capable of shaping the world after this transformation. Heidegger therefore decided to be involved—something that stems directly from his philosophy, or, to be more precise, from the concept of "involvement" central to his early philosophy.

Heidegger believed that the Nazis failed to live up to their original mission, that is, to free the West from the yoke of the dead God.[25] Seen from this perspective, that of deep history, tens of millions of people killed changes nothing. Heidegger was certain that the Nazi movement could extricate Europe and the entire West from nihilistic productive metaphysics, thereby allowing humanity to once again authentically be in the world.

Heidegger saw the Second World War as further proof that nihilism had taken over Western culture. He believed that the productive metaphysics at the root of Western culture gave rise to nihilism: "The limitless domination of modern technology in every corner of this planet is only the late consequence of a very old technical interpretation of the world, the interpretation that is usually called metaphysics" (quoted in Clark, 2011, p. 30). The Second World War was merely another step on the road to nihilism—a continuation of the fall in act from Plato to Nietzsche. As noted, to Heidegger's mind, Nietzsche—particularly in light of the "will to power" that symbolized his philosophy—did not represent liberation

[25] A reference to Nietzsche's famous pronouncement, "God is dead", expressed as follows in *The Gay Science* (1882): "The madman jumped into their midst and pierced them with his eyes. 'Whither is God?' he cried; 'I will tell you. We have killed him—you and I. All of us are his murderers. But how did we do this? How could we drink up the sea? Who gave us the sponge to wipe away the entire horizon? What were we doing when we unchained this earth from its sun? Whither is it moving now? Whither are we moving? Away from all suns? Are we not plunging continually? Backward, sideward, forward, in all directions? Is there still any up or down? Are we not straying as through an infinite nothing? Do we not feel the breath of empty space? Has it not become colder? Is not night continually closing in on us? Do we not need to light lanterns in the morning? Do we hear nothing as yet of the noise of the gravediggers who are burying God? Do we smell nothing as yet of the divine decomposition? Gods, too, decompose, God is dead. God remains dead. And we have killed him'" (Nietzsche, 1974, p. 181).

from nihilism, but its culmination, "an inextricable entanglement in metaphysics" (Heidegger, 1977, p. 61).

For Heidegger, the capitulation to technological metaphysics, to the technological way of thinking, to technological being, symbolizes the final fall of Western culture into nihilism. Hence his constant aspiration to destroy Western metaphysical thought—to save the West from nihilism. As we have seen, it is a goal he believed only the Germans, as the closest modern people to the ancient Greeks, could achieve.

Heidegger (modestly) saw himself as a philosopher capable of guiding this immense project, which required not only a philosopher but a leader. As Nancy argues (2017), Heidegger believed that Hitler was the right person at the right time: Hitler had the potential to save the West from its fall into nihilism, the ultimate expression of which is the total capitulation to technology. Hitler could have, like the Greeks, begun anew, thereby preventing the self-destruction of Western culture, but the Nazis failed. The reason they failed, from Heidegger's perspective was because they were too human, too materialistic, and not spiritual enough—trapped in the technical understanding of the world.[26]

At this stage, I think it is fair to say, based on the evidence we have seen, that Heidegger knew exactly what he was involved in. He also knew that he had no problem pressing his own philosophy into the service of Nazism, perhaps in the naïve belief that the Nazis would promote it, or perhaps convinced of a deep affinity between the essence of "ideal" Nazism and his thought.

After the Second World War, Heidegger sought to have his cake and eat it too: to clear his name and to enable a return to the essence of Nazism as he had perceived it—in its greatness. To that end, he turned to technology, in which he identified the possibility of a profound exoneration: if technology is super-metaphysics, Nazism is entirely innocent. In order to

[26] Both Heidegger and Jünger would agree that the Nazis lost because they were *too human*. For Jünger, however, who adopted the Manifesto of Futurism ("We declare that the splendor of the world has been enriched by a new beauty: the beauty of speed. A … roaring motor car which seems to run on machine-gun fire, is more beautiful than the Victory of Samothrace … We want to glorify war—the only cure for the world—militarism, patriotism, the destructive gesture of the anarchists, the beautiful ideas which kill, and contempt for woman" [in Joll, 1960, pp. 139–140]), the Nazis lost the war precisely because they did not surrender completely to the metaphysics of technology. In other words, from Jünger's perspective, the *inner truth* is the possibility of surrendering completely to the totality of technology and thereby going beyond nihilism.

attain this exoneration, Heidegger had to convince us that we are dominated by technology:

> The name "technology" is understood here in such an essential way that its meaning coincides with the term "completed metaphysics." It contains the recollection of *techne*, which is a fundamental condition of the essential development of metaphysics in general. At the same time, the name makes it possible for the planetary factor of the completion of metaphysics and its dominance to be thought without reference to historiographically demonstrable changes in nations and continents. (Heidegger, 1973, p. 93)

Four years after the war, Heidegger spoke in Bremen and presented his approach to technology. His famous assertion, "Agriculture is now a motorized food industry, the same thing in its essence as the production of corpses in the gas chambers and the extermination camps, the same thing as blockades and the reduction of countries to famine, the same thing as the manufacture of hydrogen bombs", is not an incidental remark but his final destination; it is the goal of his later philosophy: to eliminate human (German) responsibility and to make technology responsible, to argue that within the framework of nihilistic technological metaphysics, genocide and mass murder are inevitable.

Heidegger associated Nietzsche's nihilism with the nihilism to which Nazism (not the "ideal" Nazism that he had envisioned) degenerated, which is why in "Overcoming Metaphysics", written between 1936 and 1946, he quotes the notes to the fourth part of *Thus Spoke Zarathustra*: "We are attempting a venture with truth! Perhaps humanity will perish by it! So be it!" (Heidegger, 1973, p. 96). He was clever enough to put it that way and leave it to his admirers (including quite a few Jews) to do the rest. Indeed, the discourse on technology that has developed since Heidegger,[27] has accepted, to one extent or another, a considerable part of his approach, thereby subsuming—consciously or unconsciously—the Heideggerian claim that humans (Nazism) are not to blame, or at least not exclusively to blame, for the horrors that have occurred. When Heidegger links "Judaism", that is calculative thinking and productiveness, to the essence of technology, he is essentially blaming the Jews for their own destruction.

The sentence "Agriculture is now a motorized food industry, the same thing in its essence as the production of corpses in the gas chambers and the extermination camps" represents the end point in a total nihilistic

[27] For wider debate see: (Ataria & Tamir, 2023).

technological framework. It is not an incidental remark, but the very core of Heidegger's thought in the years after the war. Thus, every time we come across the word Enframing (*Ge-stell*), a fundamental concept in Heidegger's philosophy of technology, we are thrown directly to that sentence, which, I would argue, should be understood as the very essence of post-war Heideggerian thought: "A motorized food industry [is] the same thing in its essence as the production of corpses in the gas chambers and the extermination camps." Heidegger's philosophy of technology aims to bring us to accept this idea; that is its goal. We must also remember, however, that even if we accept Heidegger's approach, he fails to provide a satisfactory answer to the following question: *if what we are talking about is the technological metaphysics of domination (destruction as domination), why the Jews?*[28] I think the answer to this question is shocking beyond belief.

Returning to the possibility of a profound connection between the essence of **t**echnology and the essence of "Judaism", one might say that although Heidegger did not propose a plan for the eradication of Judaism—in an attempt to establish a free relationship between humans and **t**echnology—he did offer a path to creating a necessary and deep connection between two complete systems that can be characterized in terms of thought, practice, and belief: Judaism and **t**echnology.

As we have seen, Heidegger associated **t**echnology with Judaism—the technological mode of being is, ultimately, the Jewish mode of being. Heidegger's metaphysical anti-Semitism identifies Judaism and **t**echnology as a single metaphysical category, an identical way of thinking, acting, and being in the world, in keeping with the essence of **t**echnology, calculation, rationality, and techno-scientific logic. **T**echnology, which Heidegger blamed for the genocide of the Jewish people, is thus essentially Jewish. To support this idea, Heidegger went as far as to cite prophecy:

[28] Lyotard puts it as follows: "If we now address the interpretation that imputes the Shoah to the *Gestell*, we find that it does not answer the question of why 'the Jews' any better. As much as it can make sense of the effects that touch on economic, scientific, and technological development, on nuclear energy, agribusiness, and underdevelopment, among others (among those no less shocking, which, under the name of 'culture' touch upon language and *aisthesis*), it leaves in the dark, 'in the shadows,' that is to say, on the side, untouched, the 'reason' (since one is now under the requirement to give reasons) why 'the Jews' were chosen to prove, in all the senses of the term, that sort of 'effect'" (Lyotard, 1990, p. 88).

> "Prophecy" is the technology of warding off the element of destiny in history [*Geschicklichen der Geschichte*]. It is an instrument of the will to power. That the great prophets were Jews is a fact whose secret has not yet been thought … "Prophecy" is forward-looking history and therefore the technological perfection of the *Wesen* of history. (Quoted in Lapidot, 2020, p. 118)

Thus, according to Heidegger, Jewish metaphysics ("world-Judaism") and technological metaphysics are the culmination of nihilism and they converge both at the starting point (the Jewish prophets are technology: "'Prophecy' is the technology of warding off the element of destiny in history") and at the end point, in the form of self-destruction, which is the greatest completion of technology. From Heidegger's perspective, the Holocaust was Jewish self-destruction. It is worth noting here that this is a typically Nazi way of thinking: blaming the Jews for the Second World War. See, for example, Hitler's famous speech at the German Reichstag on 30 January 1939:

> Today I will once more be a prophet: If the international Jewish financiers in and outside Europe should succeed in plunging the nations once more into a world war, then the result will not be the Bolshevization of the earth, and thus the victory of Jewry, but the annihilation of the Jewish race in Europe! (Baynes, 1942)

The Nazis would thus appear to have been "too poorly equipped for thought to arrive at a really explicit relationship to what is happening today and has been underway for the past 300 years" (Heidegger, 1981, p. 61). Indeed, had they allowed Heidegger to be their philosophical shepherd, they would not only have annihilated the Jews, but would have been philosophically absolved of responsibility—even without destroying the evidence. From this perspective, it is hard to agree with Hannah Arendt, Heidegger's lover, who argued that he was simply "politically naïve". Heidegger seems to have been well aware of what he was getting into. He wanted—in the sense of a desire stemming from his philosophical thought—to be a part of the Nazi regime that saw Jews as a metaphysical problem that must be gotten rid of.[29]

[29] In this context, it is interesting to note Lyotard's argument in *Heidegger and "the Jews"* (1990) that one of the reasons for the Nazi annihilation of the Jews at Auschwitz was the fact that they represented the unrepresentable and, as such, undermined technological metaphysics in which representation nullifies reality. The Jews were what was preventing the West from creating a complete representation of the world, while nullifying the world itself.

In retrospect, Heidegger seems to have succeeded, if only in a limited fashion, in steering discourse away from questions of German responsibility and his own responsibility for what happened in the years 1933–1945, and toward the danger inherent to technology, as he described it, of turning human beings into a "standing-reserve", ready for reuse.

It is important to remember, however, that Heidegger, like many German intellectuals, chose the path of Nazism (Sluga, 1995). Did he fully understand what he was choosing at the time? In light of the above, it is hard to claim the Heidegger just happened to fall into Nazism. There seems to have been a decision on his part, an authentic choice—of which his rectorship address is but one piece of evidence.

In the previous section we saw how attempts have been made to absolve humans of responsibility for their actions, based on the argument that murder is in our genes. In the present chapter we have seen attempts to find exoneration in another argument: it is not the Germans but technology that is responsible for the genocide of the Jewish people, while the spirit of that technology is a fundamentally "Jewish" creation. Heidegger's sophisticated argument was that since the *essence of technology* and the *essence of Judaism* belong to the same category and self-exhaustion and self-destruction are inherent to technology, as they are to world-Judaism, it is in fact the *Jewish-technological mindset that is responsible for the genocide of the Jews*—in which case, no one is to blame for it. As we will see in the next chapter, Oppenheimer offered similar explanations for the dropping of the atomic bomb.

References

Agamben, G. (2002). *Remnants of Auschwitz: The witness and the archive. Homo Sacer III* (D. Heller-Roazen, Trans.). Zone Books.

Arendt, H., & Heidegger, M. (2004). *Letters 1925–1975* (U. Ludz, Ed., & A. Shields, Trans.). Harcourt.

Arendt, H., & Jaspers, K. (1992). *Hannah Arendt/Karl Jaspers Correspondence, 1926–1969* (L. Kohler, H. Saner, Eds., R. Kimber, & R. Kimber, Trans.). Harcourt Brace & Company.

Ataria, Y. (2016). The death of the witness in the era of testimony: Primo Levi and Georges Perec. In Y. Ataria, D. Gurevich, H. Pedaya, & Y. Neria (Eds.), *Interdisciplinary handbook of trauma and culture* (pp. 195–216). Springer.

Ataria, Y. (2017). *The structural trauma of western culture: Toward the end of humanity*. Palgrave Macmillan.

Ataria, Y. (2018). *Body Disownership in Complex Post-Traumatic Stress Disorder*. Palgrave Macmillan.

Ataria, Y. (2021). *Primo Levi and Ka-Tsetnik: The Map and the Territory*. Palgrave Macmillan.

Ataria, Y. (2022). *Consciousness in Flesh*. Palgrave Macmillan.

Ataria, Y., & Tamir, L. (2023). Why overcoming heideggerian intellectualism should precede overcoming metaphysics. *Human Studies*, 1–23.

Baynes, N. H. (Ed.). (1942). *The Speeches of Adolf Hitler, April 1922–August 1939* (Vol. 1) (B. N. H, Trans.). Oxford University Press.

Clark, T. (2011). *Martin Heidegger*. Routledge.

Derrida, J. (1989). *On Spirit: Heidegger and the Question* (G. Bennington, & R. Bowlby, Trans.). University of Chicago Press.

Di Cesare, D. (2018). *Heidegger and the Jews: The Black Notebooks* (M. Baca, Trans.). Polity Press.

Felman, S., & Laub, D. (1992). *Testimony: Crises of witnessing in literature, Psychoanalysis and history*. Routledge.

Heidegger, M. (1966). *Discourse on thinking* (J. M. Anderson, & E. H. Freund, Trans.). Harper and Row.

Heidegger, M. (1973). Overcoming metaphysics. In M. Heidegger (Ed.), *The End of Philosophy* (J. Stambaugh, Trans. pp. 84–110). Harper and Row.

Heidegger, M. (1977). *The question concerning technology and other essays* (W. Lovitt, Trans.). Harper & Row.

Heidegger, M. (1981). "only a god can save us": The Spiegel interview (1966). In T. Sheehan (Ed.), *Heidegger: The Man and the Thinker* (W. J. Richardson, Trans., pp. 45–67). Precedent Publishing.

Heidegger, M. (1982). Language in the poem. A discussion on georg Trakl's poetic work (P. D. Hertz, Trans.). In *On the way to language* (pp. 159–198). Harper & Row.

Heidegger, M. (1996). Hölderlin's hymn: "The Ister." (W. McNeill & J. Davis, Trans.). Indiana University Press. (Original work published in 1942).

Heidegger, M. (1998). *Pathmarks* (F. A. Capuzzi et al., Trans. and W. McNeill, Ed.). Cambridge University Press.

Heidegger, M. (1999). *Contributions to philosophy (From Enowning)*. (P. Emad & M. Kenneth, Trans.). Indiana Univeristy Press.

Heidegger, M. (2000). *Introduction to metaphysics* (G. Fried, & R. Polt, Trans.). Yale University Press.

Heidegger, M. (2003). The self-assertion of the German university. In M. Heidegger, & M. Stassen (Ed.), *Philosophical and political writings* (K. Harries, Trans., pp. 2–11). Continuum.

Heidegger, M. (2012). *Bremen and Freiburg lectures* (A. J. Mitchell, Trans.). Indiana University Press.

Heidegger, M. (2017). *Ponderings XII–XV: Black notebooks 1939–1941* (R. Rojcewicz, Trans.). Indiana University Press.

Heidegger, M., & Marcuse, H. (2004). Heidegger and Marcuse: A dialogue in letters. In H. Marcuse, & D. Kellner (Ed.), *Collected Papers of Herbert Marcuse, Vol. 1: Technology, War and Fascism* (W. Richard, Trans., pp. 261–267). Routledge.

Joll, J. (1960). *Three intellectuals in politics.* Pantheon Books.

Jonas, H. (1984). The imperative of responsibility. In *Search of an Ethics for the Technological Age* (H. Jonas, & D. Herr, Trans.) Chicago University Press.

Kravitz, A. (2019). Heidegger the Nazi, Heidegger the antisemite—another look (Heb.). *Tabur, 9*, 23–27.

Lang, B. (1996). *Heidegger's silence.* Cornell University Press.

Lapidot, E. (2020). Is technology Jewish? A conversation with Heidegger. *Journal of Jewish Thought and Philosophy, 28*, 95–120.

Levinas, E. (1989). As if consenting to horror. *Critical Inquiry, 15*(2), 485–488.

Löwith, K. (1993). The political implications of Heidegger's existentialism. In R. Wolin (Ed.), *The Heidegger controversy: A critical reader* (pp. 167–184). MIT Press.

Lyotard, J. F. (1990). *Heidegger and "the Jews".* (A. Michel, & M. S. Roberts, Trans.). University of Minnesota Press.

Nancy, J. L. (2017). *The banality of heidegger* (J. Fort, Trans.). Fordham University Press.

Neumann, B. (2019). *New histories of nazism* (Heb.). (A. Shapira, Ed.). Modan Publishing House.

Nietzsche, F. (1974). *The gay science* (W. Kaufmann, Trans.). Vintage Books.

Oltermann, P. (2014, March 13). Heidegger's 'Black Notebooks' reveal antisemitism at core of his philosophy. *The Guardian.* Retrieved from https://www.theguardian.com/books/2014/mar/13/martin-heidegger-black-notebooks-reveal-nazi-ideology-antisemitism

Ott, H. (1994). *Martin Heidegger: A political life* (A. Blunden, Trans.). Fontana Press.

Pimentel, D. (2019). *Heidegger with Derrida: Being Written* (N. Olshansky-Ashtr, Trans.). Palgrave Macmillan.

Pippin, R. B. (2002). Nietzsche, Heidegger, and the metaphysics of modernity. In K. Ansell-Pearson (Ed.), *Nietzsche and modern German thought* (pp. 282–310). Routledge.

Rojcewicz, R. (2006). *The gods and technology: A reading of Heidegger.* State University of New York Press.

Rose, P. L. (1990). The German moralists and the Jewish question: Kant, Herder, and Hegel. In P. L. Rose, *German question/Jewish question: Revolutionary antisemitism in Germany from Kant to Wagner* (pp. 91–116). Princeton University Press.

Sheehan, T. (1988). Heidegger and the nazis. *The New York Review of Books, 35*(10), 38–47.

Skrbina, D. (2015). *The metaphysics of technology* (1st ed.). Routledge.

Sluga, H. (1995). *Heidegger's crisis: Philosophy and politics in Nazi Germany.* Harvard University Press.

Steiner, G. (1987). *Martin Heidegger.* University of Chicago Press.

Trawny, P. (2016). Heidegger and the Shoah. In I. Farin & J. Malpas (Eds.), *Reading Heidegger's black notebooks 1931–1941* (pp. 169–179). MIT Press.

Weinberger, J. (1992). Politics and the problem of technology: An essay on Heidegger and the tradition of political philosophy. *American Political Science Review, 86*(1), 112–127.

Wolin, R. (2015). *Heidegger's children: Hannah Arendt, Karl Löwith, Hans Jonas, and Herbert Marcuse.* Princeton University Press.

Zimmerman, M. E. (1990). *Heidegger's confrontation with modernity: Technology, politics, and art.* Indiana University Press.

CHAPTER 7

Techno-Science: A Dynamic of Complete Annihilation

The Tragedy of Science at Its Peak

"A Very Wobbly Unstable Drop"

Otto Frisch and his aunt Lise Meitner, both Jewish physicists forced to leave the German Reich (Frisch in 1933 and Meitner, who lived in Germany but held Austrian citizenship, in 1938), met over Christmas 1938 in Sweden and, on a morning stroll, discussed the results of experiments conducted by one of Meitner's German colleagues, the chemist Otto Hahn, after her departure.

In a letter that Hahn sent to Meitner, he described an experiment that appeared to reveal a new physical phenomenon: "splitting the atom". For this discovery, Hahn would win the Nobel Prize in Chemistry—a prize that many believed should have gone to Meitner (Fig. 7.1). Based on their calculations, Frisch and Meitner concluded that under conditions of high surface tension, the uranium nucleus is "a very wobbly unstable drop", ready to divide at the slightest intervention. Two weeks later, in Denmark, Frisch broke the news to Niels Bohr,[1] a Jewish physicist, father of quantum mechanics, and 1922 Nobel laureate in physics, who immediately

[1] It was not a coincidence that these scientists went to Denmark. Bohr tried to find positions for Jewish physicists left jobless and vulnerable with the Nazi rise to power.

Y. Ataria, *Genes, Technology, and Apocalypse*,
https://doi.org/10.1007/978-3-031-59197-6_7

Fig. 7.1 Lise Meitner and Otto Hahn at laboratory, Kaiser-Wilhelm Institut für Chemie, Berlin (1913)

understood the ramifications and declared: "Yes, it would be possible to make a bomb, but it would take the entire efforts of a nation to do it" (in Baggott, 2010, p. 15).

The year 1939 was not the only one in which Hitler precipitated a world war that would change the face of history, but it was also the year in which the idea of splitting the atom was first published in leading scientific journals around the world. It is worth noting here that it was immediately clear to physicists that this was not merely another theoretical scientific discovery, but the basis for a new type of weapon. As Jim Baggott remarks in *The First War of Physics*, "The devastation that could be wrought with atomic weapons was obvious to the physicists right from the start" (p. xii).

Already in the early 1930s, Leo Szilard, a Hungarian-Jewish physicist, who had a nearly prophetic understanding of the geopolitical arena, conceived the chain reaction that could lead to an atomic explosion and understood the implications of publications in the field. Szilard, perhaps the most fascinating figure in the entire story of the bomb, tried to stop scientists from publishing their discoveries on the subject because he realized that they represented a formula for building an atomic bomb. Later, already in the United States, it would be Szilard and the Italian physicist Enrico Fermi, Nobel laureate for 1938, who would achieve a breakthrough in the chain reaction in uranium.

In April 1939, German scientists tried to explain to the leaders of the Reich, the importance of the discovery. For example, a 24-year-old chemist by the name of Paul Harteck wrote:

> We take the liberty of calling your attention to the newest development in nuclear physics, which, in our opinion, will probably make it possible to produce an explosive many orders of magnitude more powerful than the conventional ones … That country which first makes use of it has an unsurpassable advantage over the others. (ibid., p. 16)

In that same, fateful year, three Hungarian-Jewish scientists who had seen the actions of the Nazi regime up close, Leo Szilard, Eugene Wigner, and Edward Teller (the "Hungarian Conspiracy"), understood what it would mean for the world if Hitler were to get his hands on an atomic bomb. Together with Albert Einstein, the greatest physicist of the twentieth century, they wrote a letter to US President Roosevelt, warning him of the danger of allowing Nazi Germany to obtain such a weapon first (Fig. 7.2).

"People Must Learn to Prevent Catastrophes"

Werner Heisenberg, a student of Niels Bohr and Nobel laureate for 1932, for his development of the "uncertainty principle", was a senior figure in the Nazi effort to obtain atomic weapons. In a report he wrote, Heisenberg claimed that according to his calculations, the production of a single bomb

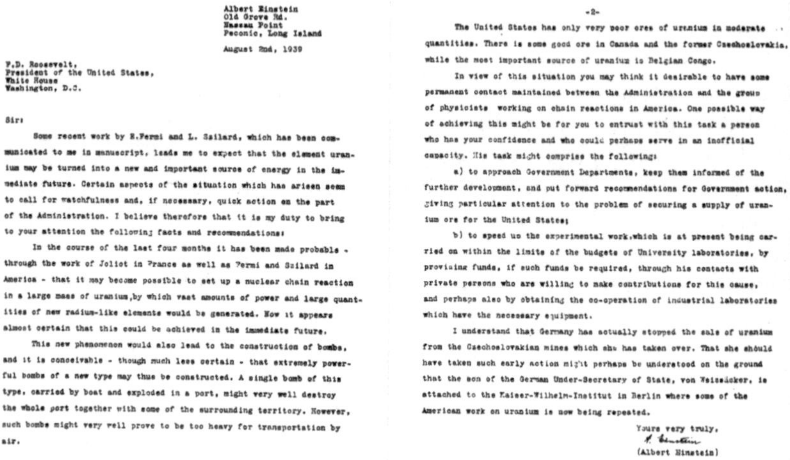

Albert Einstein
Old Grove Rd.
Nassau Point
Peconic, Long Island

August 2nd, 1939

F.D. Roosevelt,
President of the United States,
White House
Washington, D.C.

Sir:

Some recent work by E.Fermi and L. Szilard, which has been communicated to me in manuscript, leads me to expect that the element uranium may be turned into a new and important source of energy in the immediate future. Certain aspects of the situation which has arisen seem to call for watchfulness and, if necessary, quick action on the part of the Administration. I believe therefore that it is my duty to bring to your attention the following facts and recommendations:

In the course of the last four months it has been made probable - through the work of Joliot in France as well as Fermi and Szilard in America - that it may become possible to set up a nuclear chain reaction in a large mass of uranium,by which vast amounts of power and large quantities of new radium-like elements would be generated. Now it appears almost certain that this could be achieved in the immediate future.

This new phenomenon would also lead to the construction of bombs, and it is conceivable - though much less certain - that extremely powerful bombs of a new type may thus be constructed. A single bomb of this type, carried by boat and exploded in a port, might very well destroy the whole port together with some of the surrounding territory. However, such bombs might very well prove to be too heavy for transportation by air.

-2-

The United States has only very poor ores of uranium in moderate quantities. There is some good ore in Canada and the former Czechoslovakia, while the most important source of uranium is Belgian Congo.

In view of this situation you may think it desirable to have some permanent contact maintained between the Administration and the group of physicists working on chain reactions in America. One possible way of achieving this might be for you to entrust with this task a person who has your confidence and who could perhaps serve in an inofficial capacity. His task might comprise the following:

a) to approach Government Departments, keep them informed of the further development, and put forward recommendations for Government action, giving particular attention to the problem of securing a supply of uranium ore for the United States;

b) to speed up the experimental work,which is at present being carried on within the limits of the budgets of University laboratories, by providing funds, if such funds be required, through his contacts with private persons who are willing to make contributions for this cause, and perhaps also by obtaining the co-operation of industrial laboratories which have the necessary equipment.

I understand that Germany has actually stopped the sale of uranium from the Czechoslovakian mines which she has taken over. That she should have taken such early action might perhaps be understood on the ground that the son of the German Under-Secretary of State, von Weizsäcker, is attached to the Kaiser-Wilhelm-Institut in Berlin where some of the American work on uranium is now being repeated.

Yours very truly,

(Albert Einstein)

Fig. 7.2 Einstein-Roosevelt letter to Roosevelt
Opinions differ regarding Einstein's role in sending these letters, ranging from merely signing them to actually writing them. He appears to have been involved (at least) in the writing of the first letter. The same is true of two further letters sent to the US president (dated 7 March 1940 and 25 April 1940), in which Szilard asked him to accelerate development of the bomb. For an extensive discussion of Einstein's involvement, see (Schweber, 2009)

would require a massive amount of resources: more than a ton of uranium and around a ton of heavy water. Today we know that these numbers were the result of a calculation error. To this day, there is a debate regarding Heisenberg's motives. Some have suggested that the error was intentional. Secret recordings found by the Allies, however, have shown that that was not the case.[2] Moreover, we now know that Heisenberg collaborated with the Nazis, although fully aware of the dark aspects of the regime.[3] We also

[2] German scientists appear to have used moral reservations to cover up for or perhaps excuse their professional, scientific failure. Viewed in this light, Heisenberg's pointed remarks regarding Einstein's involvement in building the atom bomb (made in 1955 and in 1974, after Einstein's death) are all the more contemptible.

[3] Unfortunately, I have not come across a study that compares the attempts to defend various German figures who were deeply involved with the Nazi regime. It would be interesting to see whether there is any argumentative structure capable of exonerating Heidegger and Heisenberg (the first was politically inept and the second erred on purpose).

know that in 1937, Heisenberg was known both as a "Jew lover" and as a potential homosexual, for which reasons he was placed under Gestapo surveillance. A further fascinating point, which I would argue is far more than a simple anecdote, is that Heisenberg visited the United States in 1939 and turned down a position at Columbia University—a position that had been offered to him before, in 1937.

Heisenberg is best known for his uncertainty principle: "It's not that the electron has a position and that we don't know the position before we do our measurement. Rather, contrary to what you'd expect, the electron *simply does not have* a definite position before the measurement is taken" (Greene, 2004, p. 94). According to quantum theory, "You can determine with precision certain physical features of the microscopic realm, but in so doing you eliminate the possibility of precisely determining certain other, complementary features" (ibid., p. 96). As one of the leading physicists in the field of string theory, Brian Greene emphasizes, "The uncertainty principle is completely general*: it applies to everything*" (ibid., p. 97). It thus follows that the uncertainty principle also applies to Heisenberg himself.

During his visit to the United States, Fermi asked Heisenberg why he had turned down the offer from Columbia. Heisenberg replied, "people must learn to prevent catastrophes, not to run away from them" (Baggott, 2010, p. 27). This answer could be seen as stemming from a solid moral position and a willingness to struggle against injustice—which may be what led some to suggest that the calculation error mentioned above was intentional. The problem is that Heisenberg's actions did not match his words. With this in mind, we should at least consider the possibility that perhaps the catastrophe he was referring to was the defeat of Nazi Germany in the war. This would have been in keeping with the fact that Heisenberg later agreed "eagerly and without hesitation" to play a leading role in the Nazi atomic weapons project, even inverting the official slogan of the Nazi government, "We must make use of physics for warfare", to "We must make use of warfare for physics" (ibid., p. 28). Of course, at this point, the question arises: are scientists really so politically naïve, so detached from reality, or do they simply understand that war offers an opportunity to obtain generous funding for research, murderous consequences be damned.

In 1941, when Heisenberg was convinced that Germany was ahead of the allies in the race for the bomb, he insisted on visiting his revered teacher Niels Bohr, in order to consult with him (or so he claimed after the war) about the morality of developing an atomic weapon. Heisenberg wrote to his wife, disappointedly, that "even a great man like Bohr cannot separate out thinking, feeling, and hating entirely" (ibid., p. 89). Bohr probably understood, however, that when it comes to the atom, such separation, the attempt to think in a purely rational manner, is the most dangerous thing of all.[4]

Fascinatingly, but not surprisingly, Heisenberg and Bohr describe the meetings between them in completely different ways: the teacher remembered a student who was eager to create an atomic bomb, while the student recalled trying to reach an understanding regarding a commitment by physicists not to develop such weapons. Heisenberg even claimed, because he was under Gestapo surveillance, that he had merely hinted at the subject to Bohr. The fact remains that at the end of the meeting, Bohr was convinced that the Nazis were on the fast track to creating a bomb. Even today, "What was said or not said between Bohr and Heisenberg remains a source of considerable controversy" (Bird & Sherwin, 2005, p. 271). As we will see, Bohr's foreboding feeling would have a profound effect on the progress of the Manhattan Project (even without a direct contribution), thereby conditioning the consciences of a number of leading physicists working on creating an atomic bomb.

The Complementarity Principle

The Manhattan Project—codename for the American nuclear weapons program—was officially launched on 13 August 1942, following the Einstein-Szilard letter to President Roosevelt, recommending the United States develop a nuclear fission bomb. Working on the project were some of the twentieth century's greatest scientists, many of them Jewish Nobel laureates.

The project was headed by Robert Oppenheimer (who will be discussed at length in the second half of the chapter). Although two German-Jewish physicists, Otto Frisch and Rudolf Peierls (who were working for the British team, in collaboration with the American team), had already worked out how to build an atomic bomb and issued a

[4] Interestingly, Heisenberg himself added, "But probably one ought not to separate these ever" (in Baggott, 2010, p. 89).

report on "the possibility of constructing a 'super-bomb'" as early as the spring of 1941, the work would drag on for at least another year. It was only in the second half of 1943 that the Manhattan Project would begin to accelerate. Thus, on 30 December 1943, when Bohr came to Los Alamos, he found that the Manhattan Project was well on its way to developing an atomic bomb. As we have seen, he believed the Germans were on exactly the same track. Bohr was not troubled by technical questions, but by the fact that "the British and American administrations had given little—if any—thought to the post-war challenges posed by the threat of the weapons that would soon be available" (Baggott, 2010, p. 230).

In relation to the atomic bomb, Bohr's point of departure was based on the "complementarity principle", one of the pillars of quantum philosophy: "Every electron, every photon, *everything*, in fact, has both wavelike and particlelike aspects. They are complementary features. Thinking purely in the conventional particle framework—in which particles move along single, unique trajectories—is incomplete, because it misses the wavelike aspects demonstrated by interference patterns" (Greene, 2004, p. 185). Bohr was of the opinion that while in certain circumstances, the atomic bomb signified the end of humankind, in other circumstances, "the advent of atomic weapons meant that recourse to war would no longer be thinkable" (Baggott, 2010, p. 230). Bohr believed that in order to arrive at the second option, that is, the atomic bomb as a guarantee against future war, knowledge had to be shared with the Soviet Union and monitored together (ideas that appear to have had a profound impact on Oppenheimer). In the report he produced in the spring of 1944, apparently after having understood that the Germans would not succeed in building a bomb, Bohr wrote that, "in the coming atomic era, humanity would not be safe unless secrecy was banished". Otherwise, Bohr predicted, we would enter an endless arms race: "Unless, indeed, some agreement about the control of the use of the new active materials can be obtained in due time, any temporary advantage, however great, may be outweighed by a perpetual menace to human security" (Bird & Sherwin, 2005, p. 273).

Bohr sought to convey this message both to Roosevelt and to British Prime Minister Winston Churchill. Roosevelt appears to have been somewhat sympathetic initially, but ultimately sent Bohr to Churchill, who rejected his argument outright and even convinced Roosevelt that it was entirely without merit.[5] Churchill felt that Bohr should be locked up. Israeli historian Avner Cohen sees Bohr as "perhaps the most pathetic example of the 'private' politics conducted by a number the project's scientists in the matter of the bomb" (1989, p. 33). Indeed, Bohr not only embodies the ongoing failure of a number of scientists to explain themselves to decision-makers, but also the poor grasp these scientists had of the political sphere and the forces that operate within it.

At the meeting between Churchill and Bohr, Churchill insisted: "After all this new bomb is just going to be bigger than our present bombs", but "It involves no difference in the principles of war" (Baggott, 2010, p. 237). This statement shows how Churchill failed to fully understand the nature of atomic weapons,[6] but Bohr too failed to grasp the true nature of this type of weapon. There was never any way that atomic weapons could

[5] Geoffrey Best (2005, p. 210) argues that the fact that Bohr met with Churchill on 16 May 1944, at the height of preparations for D-Day, cannot excuse Churchill's incivility.

[6] Geoffrey Best (2005) shows how Churchill devoted little time to the atomic question during the war. He seems to have believed that it was a project beyond Britain's wartime capabilities. Furthermore, Churchill, like Truman, felt that the gap between the United States and the Soviet Union would not be bridged quickly. From a broader perspective, however, Churchill appears to have had a profound understanding of atomic weapons. As early as 1924, he published a short text titled "Shall We Commit Suicide?" in which he wondered, "May there not be methods of using explosive energy incomparably more intense than anything heretofore discovered? Might not a bomb no bigger than an orange be found to possess a secret power to destroy a whole block of buildings—nay to concentrate the force of a thousand tons of cordite and blast a township at a stroke?" In another article, published in 1931, he wrote, "What is lacking is the match to set the bonfire alight, or it may be the detonator to cause the dynamite to explode. The Scientists are looking for this" (quoted in Best, 2005, pp. 202–203). During his second term as prime minister (1951–1955), he seems to have begun to grasp the true significance of the bomb and that there was no chance that the Soviets (or the British) would not seek to obtain it. Like many others, he tried to understand whether the atom could be used to make a deterrent weapon that would prevent the Cold War from becoming a hot war, in the belief that the leading Powers "might do something better than tear the human race, including themselves, into bits" (speech to the House of Commons, 11 May 1953, quoted in Best, 2005, p. 231). In his final speech to the Commons, on 3 November 1953, Churchill said: "Indeed, I have sometimes the odd thought that the annihilating character of these agencies may bring an utterly unforeseeable security to mankind … When I was a schoolboy I was not good at arithmetic, but I have since heard it said that certain mathematical quantities when they pass through infinity change their signs from plus to minus—or the other way round … It may be that this rule may have a novel application and that when the advance of destructive weapons enables everyone to kill everybody else nobody will want to kill anyone at all" (Best, 2005, p. 236).

serve as a tool to prevent war (this would also be Oppenheimer's mistake)—certainly not in 1945. Moreover, Bohr's theory that sharing knowledge with the Soviet Union would prevent an arms race is also entirely without basis.

Someone who really understood the danger posed by the atom bomb was, once again, Szilard, who insisted on meeting with Oppenheimer. The meeting took place on 30 May 1945, a month and a half before the first nuclear test and one day before the first meeting of the Interim Committee on nuclear issues, when Oppenheimer was on his way to meet US Secretary of War Henry Stimson. Szilard did not believe that merely informing the Soviets about the bomb and the intention to use it would suffice, since the problem was not that they would fail to get the message but, on the contrary, "They'll understand it only too well" (Bird & Sherwin, 2005, p. 293). Indeed, when Truman would finally share the information with the Russian leader Joseph Stalin (information he already had), it would prove to be too little too late.

It seems that when it comes to atomic weapons, previous theories do not apply. Atomic weapons have no complementary features and are not ambiguous. Nuclear reactors may be used to produce clean energy but that merely disguises the one and only purpose of nuclear weapons: mass destruction. Nuclear reactors have been put forward as a solution to climate change, but the fact remains that nuclear energy is, ultimately, a destructive force. Chernobyl and Fukushima have shown that even nuclear reactors for non-military purposes have the potential to destroy, and since mistakes always happen, so will accidents—and in the case of nuclear reactors, accidents can be fatal.

In any event, Bohr was not thinking of nuclear reactors as a source of clean electricity, but of the idea that nuclear weapons are so powerful that they might bring about peace, due to the effect of mutual deterrence. He made a fundamental mistake, however. First, in order for a weapon to deter its own use, there would have to be a real possibility that it would actually be used, also implying real intention, and where there is real intention, it is very hard to predict how things will end. Second, cost differential will always favor offensive over defensive weapons, and once a country has primarily offensive power, the temptation to land the first blow increases, while the relative lack of defensive power is an incentive to make the first blow "count". Third, when a country decides to build a bomb, the message it is sending is crystal clear: the threat of destruction. Any other interpretation is needlessly complicated philosophy, when the bomb is a fairly straightforward story: a weapon of mass destruction. Bohr did not fully understand this—nor did Oppenheimer, whose position was significantly influenced by his conversations with Bohr over the course of 1944. The idea that the atomic bomb can be considered an object by means of which peace accords can be imposed or a mechanism for ensuring global equilibrium is thus fundamentally flawed.[7] The failure of the immense intellects to understand the horrific significance of atomic weapons stems from the attempt to apply theories developed in the pre-atomic age to the atomic age. Einstein perhaps put it best, in May 1946: "The unleashed power of the atom has changed everything save our modes of thinking and we thus drift toward unparalleled catastrophe".

When Germany was defeated, its scientists were far from building an atomic bomb. We now know that by late 1943, Allied intelligence reports already indicated that the Germans were no real competition for the Manhattan Project. The Soviets were not close to building a bomb either. The Manhattan Project was created in order to defeat the Germans, but since the Germans surrendered before one was built, the question arises: why continue developing the bomb?

Attempts to Stop

When we talk about attempts to stop the bomb from being used, the name Leo Szilard—who had the deepest political awareness of all his colleagues—stands out above all others. In 1939, Szilard recruited Einstein to join him in writing their famous letter to Roosevelt, leading to the creation of the Manhattan Project (after the second, more determined letter of 1940). After the Germans were defeated, the Manhattan Project continued and Szilard again approached Einstein to help him reach the president—this time to stop the bomb.[8] Although the letter was written and sent, Roosevelt died in early April, and the attempt failed. It would be optimistic bordering on childish to think that a letter would have stopped the project from moving forward, or that had Roosevelt lived and the meeting taken place,

[7] For an extensive discussion, see (Daley, 2010).

[8] This is generally considered Einstein's third letter, but it is in fact merely a formal introduction in which Einstein states that he knows nothing about the Manhattan Project (although he did know) and that Szilard had told him nothing (in order not to implicate him in revealing confidential information) but is simply serving as a kind of advocate on Szilard's behalf (Schweber, 2009).

Szilard would have convinced him to stop developing the bomb. Today we know that "the qualms of the scientists working on the Manhattan Project had only a very limited effect, miniscule in fact, on the thinking of decision-makers in Washington, while the latter still did not grasp the significance of entering the nuclear age" (Cohen, 1989, p. 42).

Over the course of 1945, Oppenheimer argued that "if atomic weapons were to foreshadow the end of all conventional war ... the world needed first to see the effects of these new weapons if they were to be properly respected" (Baggott, 2010, p. 300). Oppenheimer did not respond to Szilard's letter questioning this theory. Szilard was convinced that the atomic bomb would not lead to peace, but to an arms race between powers: "If a race in the production of atomic bombs should become unavoidable," he wrote, "the prospects of this country cannot be expected to be good", adding that he doubted "whether it is wise to show our hand by using atomic bombs against Japan" (Bird & Sherwin, 2005, pp. 291–292).

In order to address further nuclear issues, Henry Stimson established the Interim Committee, with the blessing of the new president, Harry Truman. Stimson, who was secretary of war at the time and largely responsible for the development of the atomic bomb and the decision to use it, was very troubled by the power of the bomb and the ramifications of using it. The Interim Committee was supposed to discuss the core issues pertaining to the use of atomic weapons. Military commanders and scientists also took part in the committee's meetings. The question of whether the bomb should be used against Japan was not on the agenda of the committee's first meeting on 31 May 1945. The subject was raised spontaneously and informally during lunch. Stimson declared that he did not regard the bomb "as a new weapon merely but as a revolutionary change in the relations of man to the universe". Under certain circumstances, he argued, it might become "a Frankenstein which would eat us up" (Bird & Sherwin, 2005, p. 293). Oppenheimer proposed civilian use of atomic energy and a free exchange of information, since the level of knowledge is such that little is required to build a bomb. Remember that by that time, most of the theoretical work had already been done. Stimson, however, was against the idea. When it comes to the Soviet Union, he argued, it is a problem, since totalitarian regimes can't be trusted. Somewhat surprisingly, US Army Chief of Staff General George Marshall actually defended the Russians, saying he was certain that "we need have no fear that the Russians, if they had knowledge of the project, would disclose this information to the Japanese" (p. 295). Secretary of State James Byrnes held the view that the atomic bomb could be used as a diplomatic weapon against the Russians. It is clear to us today just how much this position shows Byrnes's complete failure to understand the bomb, let alone the dynamics of the day after the war, that is, the arms race. It is worth noting that the minutes of the meeting make no mention of any moral objections. The committee's working conclusions were "that we could not give the Japanese any warning; that we could not concentrate on a civilian area; but that we should seek to make a profound psychological impression on as many of the inhabitants as possible" (p. 296).

Although Oppenheimer did raise objections at the meeting, he ultimately agreed with the decision-makers and took the position that "the war should not end without the world knowing about the bomb and its effects" (Baggott, 2010, p. 306).[9]

The project was centered in Los Alamos, New Mexico, but some of the scientists working on it in Chicago were very disturbed by the course of events.[10] With Szilard in the background, a number of scientists wrote a sober report (known as the Franck Report), which they submitted to the Interim Committee. At a distance of more than 70 years, it is interesting to think what might have happened had it been adopted:

> The development of nuclear power not only constitutes an important addition to the technological and military power of the United States, but also creates grave political and economic problems for the future of this country. Nuclear bombs cannot possibly remain a "secret weapon" at the exclusive disposal of this country, for more than a few years. The scientific facts on which their construction is based are well known to scientists of other countries. *Unless an effective international control of nuclear explosives is instituted, a race of nuclear armaments is certain to ensue following the first revelation of our possession of nuclear weapons to the world.* (quoted in Lemmerich, 2011, p. 325).

[9] It is hard to exaggerate Oppenheimer's importance as a mediator between the military and political leadership and the scientists. Indeed, Manhattan Project director Leslie Groves fully understood Oppenheimer's position not only at this juncture but throughout.

[10] It is worth noting that the Chicago group had already completed its role in the project by this point and may thus have been freer to express its views.

> All that these advantages can give us, is the accumulation of a larger number of bigger and better atomic bombs … *However, such a quantitative advantage in reserves of bottled destructive power will not make us safe from sudden attack.* Just because a potential enemy will be afraid of being "outnumbered and outgunned," the temptation for him may be overwhelming to attempt a sudden unprovoked blow—particularly if he would suspect us of harboring aggressive intentions against his security or "sphere of influence." *In no other type of warfare does the advantage lie so heavily with the aggressor* … From this point of view a demonstration of the new weapon may best be made before the eyes of representatives of all United Nations, on the desert or a barren island … If the United States would be the first to release this new means of indiscriminate destruction upon mankind, she would sacrifice public support throughout the world, precipitate the race of armaments, and prejudice the possibility of reaching an international agreement on the future control of such weapons. (p. 321, 323, 326)
>
> A nation which was capable of secretly preparing and suddenly releasing a weapon, as indiscriminate as the rocket bomb and a {million?}/thousand/ times more destructive, is to be trusted in its proclaimed desire of having such weapons abolished by international agreement. (p. 323)

The report did not make it into the hands of Interim Committee Chairman Henry Stimson, but into those of his personal secretary. It was delivered to Stimson's office by American physicist and 1927 Nobel laureate in physics Arthur Compton, who was president of MIT and one of the scientific representatives to the Interim Committee. Compton added a cover note, in which he pointed out that although the report did call attention to "difficulties that might result from the use of the bomb", it failed to mention "the probable net saving of many lives", or "that if the bomb were not used in the present war the world would have no adequate warning as to what was to be expected if war should break out again" (Compton, 1956, p. 236).

The Franck Report was filed as classified information and never made it to Harry Truman, who was president by that time. The scientific board that advised the Interim Committee wrote a report in which it did note that there were internal differences among the scientists, but concluded unequivocally: "We can propose no technical demonstration likely to bring an end to the war; we see no acceptable alternative to direct military use" (Baggott, 2010, p. 309). The closing lines of the report are shocking:

> It is true that we are among the few citizens who have had occasion to give thoughtful consideration to these problems during the past few years. We have, however, no claim to special competence in solving the political, social, and military problems which are presented by the advent of atomic power. (ibid., pp. 309–310)

What we see here is the epitome of technicians' morality—limiting engagement with such a momentous topic to technical problems, without the ability, or perhaps the will to consider the core issues. This morality is typical of thinking that imposes a clear separation between the technical process and its result.

Szilard, who, together with Einstein, wrote two letters to expedite the Manhattan Project, now fought with all his might against the project. He got 59 scientists from the laboratory in Chicago to sign a petition, which he sent to Oppenheimer—to which Oppenheimer's reaction remains unknown to this day. That same month, Szilard drafted another petition, signed by 68 scientists, and sent it directly to Truman. In this petition, he wrote:

> Attacks by atomic bombs may very well be an effective method of warfare. We feel, however, that such attacks on Japan could not be justified, at least not until the terms which will be imposed after the war on Japan were made public in detail and Japan were given an opportunity to surrender … *Such a step, however, ought not to be made at any time without seriously considering the moral responsibilities which are involved.* (ibid., p. 327)

The petition was passed from hand to hand, delayed at every stage. On 1 August, two weeks after the bomb was successfully tested, it finally made it to the office of Secretary of War Henry Stimson, who was, at that time, in Potsdam, Germany. He could only read it when he returned, by which time it was too late.

Development of the Hydrogen Bomb

The fact that leading scientists, such as Szilard, Compton, Bethe, Einstein, Oppenheimer, and others had woken up to the dangers of the atom bomb and its use, did not prevent the development of the hydrogen bomb. Some of the scientists invited to work on the hydrogen bomb had worked on the atomic bomb and opposed the development of the new bomb. The most famous of these was Hans Bethe, 1967 Nobel laureate in physics, who, despite his opposition to the development of the hydrogen bomb, ultimately went to work on this project as well, when his efforts failed to prevent it.

The American administration became anxious, after the Russians tested a bomb (codenamed Joe 1)—an exact copy of the American, plutonium bomb—in August 1949. At first, Truman refused to believe this development, but it came as no surprise to some of the scientists at Los Alamos. Oppenheimer called it "the fulfillment of an expectation" (Bird & Sherwin, 2005, p. 417). In response to this test, the American stockpile of atomic weapons, which numbered 50 bombs in 1948, grew to 300 bombs by 1950. But it was not enough. Truman was fascinated by the possibility of building a hydrogen bomb—a weapon for which any target would be small.

At this stage, the scientists lost their naïveté. Compton put it most bluntly: "We feel that this development [the H-bomb] should not be undertaken, primarily because *we should prefer defeat in war to victory obtained at the expense of the enormous human disaster that would be caused by its determined use*" (Bird & Sherwin, 2005, p. 418). Nevertheless, scientists were found to develop the new bomb—notably Edward Teller, known as the "father of the H-bomb", who was, as noted earlier, part of the "Hungarian Conspiracy".

Discussion of the Hydrogen bomb appears to have been intensified by fear of the Russian bomb, which, although not yet developed, would, according to game-theory logic, eventually be built and must therefore be obtained first:

> In the late 1950s, the fear of losing technological superiority to the Russians gave rise to the myth of a Russian "missile gap" … The American Central Intelligence Agency gave public voice, at the time, to pessimistic remarks on which the myth was based … influencing the decisions of the Kennedy administration in matters of security. A short time after these assessments were made public, however, it became clear that they were entirely without basis … fifteen times less than the estimates publicized by the [same] Intelligence Agency a few years earlier … By that time, it was too late to stop the decision to develop 1000 intercontinental missiles … [and in another case,] the fears concerned air defence systems that the Soviets had begun to consider developing. (Cohen, 1989, pp. 97–115)

We now know that, with the exception of a very brief period, the Soviet Union did in fact lag behind the United States. 1952 saw the "birth" of Mike—a hydrogen bomb 100 times more powerful than Little Boy.

Was It Really an Inevitable Dynamic?

Baggott argues that "This confluence of historical events", that is, the discovery of the potential of the atom in 1939, with the Second World War was fast approaching, "meant that there was a *startling inevitability* about the chain that led from the discovery of fission to the development of the bomb, to the use of the bomb against Japan, to the development of a Soviet weapon" (Baggott, 2010, p. 462). If so, it is a matter of historical determinism.

At this point, however, it is worth going back to Lise Meitner, dubbed the "Jewish mother of the bomb". When Meitner was approached to join the Manhattan Project, her answer was clear and unequivocal: "I will have nothing to do with a bomb" (Baggott, 2010, p. 217).

Meitner's refusal proves that another path might have been possible, had people only learned to say "no", to refuse to take part. Her refusal raises profound questions about the idea of an inevitable techno-scientific dynamic, on the path to total annihilation. In light of this, let us now take a look at the figure of Robert Oppenheimer, also known as the "father of the atomic bomb".

Robert Oppenheimer: Technical Sweetness

Who Was Robert Oppenheimer?

Robert Oppenheimer (Fig. 7.3), a Jewish-American physicist whose parents had immigrated to the United States from Germany in the late nineteenth century, completed his PhD in Germany, at the age of 23. He was an intellectual, brilliant, cultured, arrogant, elitist, politically involved (at least since the Great Depression), and the head of the US atomic project (Manhattan Project) who, within the space of 2 years, led a team of scientists to create an atomic bomb. I will not go into the nature of his association with the Communist movement in the 1930s, but the fact that he was involved, albeit not as a party member, attests to a man who was active in politics and who contributed money, energy, and time to a variety of social causes (such as the farm protests, civil rights organizations, and the Spanish Republican government). He was by no means detached from reality or lacking in political awareness. It is important to start from this point, since it teaches us something fundamental about him.

Fig. 7.3 Robert Oppenheimer—1944

Oppenheimer was not a man who was easily swayed, and, as a star in the world of physics, it is unlikely that he was after fame or glory. Although one can never have enough glory and will always strive for further recognition, to say that Oppenheimer did what he did out of arrogance or a desire for glory (alone) would be rather superficial. Oppenheimer—who was, as everyone who knew him agreed, an independent-minded person, firm in his convictions—was recruited to the Manhattan Project with his eyes wide open and, in fact, did everything in his power to join it.

Was an Atomic Bomb the Only Way to Dislodge Hitler from Europe?

In late 1941, Oppenheimer was asked to help with the calculations to determine the amount of uranium necessary to create a nuclear bomb. In January 1942, when he discovered that he was a candidate to head the fast neutron research group at Berkeley, he distanced himself from any political activity that could have hurt his chances, and even made this known to the relevant parties. In May 1942, he received the appointment, with the title Coordinator of Rapid Rupture.

Oppenheimer organized a summer seminar with the goal of laying the groundwork for the development of an atomic bomb. He recruited physicists, including Teller (part of the "Hungarian Conspiracy") and Hans Bethe—a German-Jewish physicist who had fled Germany after Hitler's rise to power. During the course of the seminar, Oppenheimer proved to be not only a gifted physicist, but also a leader and someone who could stick to a timetable. In the words of Bethe: "Now I could see at firsthand the tremendous intellectual power of Oppenheimer who was the unquestioned leader of our group" (Bird & Sherwin, 2005, p. 182).[11]

In order to gain further insight into Oppenheimer's frame of mind and that of the other scientists who took part in the summer seminar, let us return to Teller, who recalled Oppenheimer saying to him one evening that "only an atomic bomb could dislodge Hitler from Europe" (ibid., p. 183).

Despite his successes, the authorities refused to give Oppenheimer the security clearance that would have allowed him to take part in the nuclear project, due to his communist past. Oppenheimer did everything he could to obtain the desired clearance. His inability to let it go would cost him dearly toward the end of his life.

On 18 September 1942, Oppenheimer met with Leslie Groves, head of the military project, and managed to convince him of his abilities and importance. There was probably more to his success in convincing Groves than his brilliance, knowledge, tremendous intellect, phenomenal memory, or the fact that Groves believed him to be one of the few people who understood the need for applied solutions.[12] Some have suggested that more than anything else, Groves was impressed by his personal ambition.[13]

[11] Bethe appears to be one of the few people to remain loyal to Oppenheimer to the end, continuing to support him despite the criticism against him. Among other things, he tried to dissuade Teller from testifying against him during the 1953–1954 hearings.

[12] This is a little odd, considering the fact that Oppenheimer was very weak in applied physics as a student.

[13] Some have suggested that Groves also identified in Oppenheimer a need to satisfy his superiors, which is why he chose him to lead the project (Schweber, 2000). Monk, on the other hand, chose to stress Oppenheimer's arrogance (Monk, 2012). Personally, I have my reservations about this type of psychologizing discourse, which reduces historical processes to the personality traits of this or that individual. There is no real difference between reducing things to personality and reducing them to genetics, for example. In this sense, psychoanalytical discourse is particularly problematic, due to its extreme reductionist tendencies.

Groves and Oppenheimer (1942)

"What If the Nazis Get It first?"

As soon as Oppenheimer was appointed scientific director of the project, he began to recruit leading scientists, and to those skeptics who asked whether he truly understood the implications of developing the bomb, he simply replied, "What if the Nazis get it first?" (ibid., p. 188).

At first, Oppenheimer had administrative difficulties, but he quickly learned his role and became a charismatic leader who led the project and swept the others after him, navigating between impossible military regulations, expanding the horizons of scientific thought, and finding practical solutions to complex problems.

One of the scientists that Oppenheimer tried to recruit was Isidor Rabi, a Jewish-American physicist and Nobel laureate for 1944, but Rabi refused. He was working on the radar project, which was seen as far more relevant and practical at that time, but he also refused to work on the atomic bomb on moral grounds. He expressed fear that "the culmination of three centuries of physics" would be a weapon of mass destruction. Oppenheimer once again replied, "I do not think that the Nazis allow us the option of [not] carrying out that development" (ibid., p. 212). In retrospect, Rabi seems to have internalized the true significance of the bomb long before Oppenheimer.

In early 1945, when it was clear that the Nazis, with or without the bomb, were facing defeat, conversation among the scientists in Los Alamos regarding the need for the bomb became more heated. Oppenheimer remained steadfast in his views and urged the scientists to continue to devote all of their energy to the project. In internal discussions the physicists conducted on the necessity and morality of the project, Oppenheimer presented two main arguments: first, that the scientists working on the project are merely technicians and have no particular say in when, how, or why the bomb should be used; second, following Bohr, that the bomb could put an end all wars, and in order to achieve that goal, the war must not end without the world knowing about it.

"This Is a Weapon Which Has No Military Significance"

It is not always easy to follow Oppenheimer's logic. For example, at a meeting with Szilard in May 1945, a day before the Interim Committee meeting, Szilard expressed his concerns about the bomb, to which Oppenheimer replied: "Well, this is a weapon which has no military significance. It will make a big bang—a very big bang—but it is not a weapon which is useful in war" (ibid., p. 292). It is not entirely clear what Oppenheimer meant by this. Was it naïveté or some deep insight into the nature of a weapon that was beyond war—one that, in a certain sense, cannot even be called a weapon. Indeed this remains a fundamental question to this day: in what sense are atomic weapons really weapons?

Oppenheimer did not only head the scientific project. He was also the scientists' representative on various committees, and it must be said: although he did try to present a more complex position on occasion—at the first meeting of the Interim Committee, on 31 May 1945, for example—ultimately, when his position regarding openness and the sharing of information was rejected, he "acquiesced to everything" (p. 297). That is to say that the Soviets would be given no significant advance briefing on the bomb and its ramifications, and it would be dropped on Japan without warning. In other words, despite his intentions, Oppenheimer, a man of immense intellect who understood and was involved in politics, failed to present a real alternative. Or perhaps he was not interested in doing so.

Oppenheimer's Lowest Point

To my mind, Oppenheimer's lowest point was the report he submitted to the Interim Committee on 16 June 1945. Although he knew that many scientists in Los Alamos and in Chicago supported only a demonstration of the bomb, and was well aware of the differences of opinion among them, he chose to present the message that the use of the bomb could prevent all future wars and that there was no alternative to its direct military use.

On the one hand, Oppenheimer seems to present a solid position, worthy of respect even if one does not agree with it, but his self-effacement in the final part of the report, the clear assertion that the scientists who worked on the project had no special right or competence to judge or decide on the matter, attests to an intellectual and moral weakness at a stage when he might have changed the course of the bomb in human history.

At the moment of truth, Oppenheimer chose a tortuous and problematic path: he argued that the bomb should be used and that it could lead (under certain circumstances) to peace, but, at the same time, declared that his opinion should not be paid too much attention. Oppenheimer fled from responsibility for his actions.

It is worth noting that Oppenheimer was not only the scientific director of the Manhattan Project, but also, due to his role, had a place of honor at the decision-making table. In this context we must ask why, if his opinion was not important, did he insist on voicing it, time and again? In practice, he thwarted the opposition to the use of the bomb expressed by other scientists on the project.

I call this fleeing from responsibility and not momentary weakness, since, in June 1945, 2 weeks after he had submitted the scientists' report to the Interim Committee, he was presented with Szilard's petition calling on President Truman not to use the nuclear bomb "unless the terms which will be imposed upon Japan have been made public in detail and Japan knowing these terms has refused to surrender." It is important to recall that 155 scientists signed this petition, and when Teller presented it to Oppenheimer, the latter was evasive and said that the decisions should be left up to those who know what they are doing, like Stimson and Marshall.

"Now I Am Become Death, the Destroyer of Worlds"

After the first test of the atomic bomb, Trinity, on 16 July 1945 (Figs. 7.4 and 7.5), Rabi said that he noticed Oppenheimer's gait: "I'll never forget his walk; I'll never forget the way he stepped out of the car ... his walk was

Fig. 7.4 Trinity—the first nuclear test explosion (July 16, 1945)

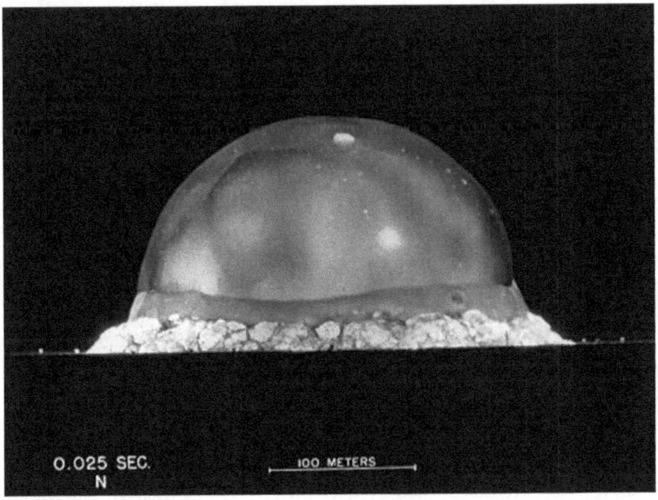

Fig. 7.5 Trinity Test Fireball 25ms

like High Noon … this kind of strut" (Bird & Sherwin, 2005, p. 308). Oppenheimer later said that when he saw the mushroom cloud in the sky, he was reminded of a line from the Bhagavad-Gita: "'Now I am become death, the destroyer of worlds'. I suppose we all thought that, one way or another" (ibid., p. 309). A few days after the test, but before Hiroshima, Oppenheimer was troubled. Robert Wilson, one of the project's leading scientists, who was concerned about using the bomb and encouraged open debate on the subject, recalled that Oppenheimer said to him, "Those poor little people, those poor little people" (ibid., p. 314). In light of this, we might have expected to find a more hesitant Oppenheimer in the following days, yet, at a meeting with the man directly responsible for the bombing of Hiroshima, Oppenheimer insisted on every detail, in order to have the greatest effect and inflict the most damage:

> There should be a moon; that would be best. Of course, they must not drop it in rain or fog … Don't let them detonate it too high. The figure fixed on is just right. Don't let it go up [higher] or the target won't get as much damage. (ibid., p. 314)

On 6 August, at two in the afternoon, Groves called Oppenheimer to tell him about the great success. Oppenheimer replied, "Everybody is feeling reasonably good about it". In the evening, the scientists gathered in an auditorium and cheered as Oppenheimer, on the stage, "clasped his hands together and pumped them over his head like a prizefighter," saying, among other things that it was "too early to determine what the results of the bombing might have been, but he was sure that the Japanese didn't like it" (ibid., pp. 315–316).

"I Feel I Have Blood on My Hands"

Following the atomic explosion in the skies above Japan, Oppenheimer became a star, although it must be said, to his credit, that he used his standing immediately after the bomb was detonated to preach against the bomb that he himself had created: "The atomic bomb is by definition a weapon of terror and aggression ... The pattern of the use of atomic weapons was set at Hiroshima ... against an essentially defeated enemy" (ibid., pp. 323–324). Even after his resignation as scientific director of the Manhattan Project in late 1945, he said, at a farewell ceremony: "If atomic bombs are to be added as new weapons to the arsenals of a warring world, or to the arsenals of nations preparing for war, then the time will come when mankind will curse the names of Los Alamos and Hiroshima" (ibid., p. 329).

In contrast, on 25 October 1945, he met with President Harry Truman in order to promote his plan for international supervision over the bomb. Truman asked him to support a law that would give the military full control in all matters concerning nuclear energy. Oppenheimer, who could never suffer stupidity, fell silent after Truman asked him when he thought the Russians would get a bomb of their own. In light of Oppenheimer's reply that *he did not know*, Truman himself answered, "Never!" From Oppenheimer's perspective, this answer showed a complete lack of understanding of the dynamic of nuclear weapons. Yet, in the debate over the hydrogen bomb, Oppenheimer himself proved not to understand the dynamic of the atom, embodied in the simple slogan: "If it can be done, it should be done".

By this stage, Oppenheimer would have had no doubts about who Truman really was. The man who oversaw the last part of the Second World War should have understood the power of the arsenal he had at his disposal, but seems to have failed to grasp what exactly an atomic bomb is—its power, its nature, and, especially, how impossible it was to keep it a secret. It is safe to assume that Oppenheimer realized that he had in fact been deceiving himself and the scientists in Los Alamos when he repeated again and again that the administration could be trusted, that is to say that the decision-makers knew what they were doing and made the right decisions based on a clear understanding of the nature of the bomb and the exigencies of the war. Worse yet, Oppenheimer almost certainly knew that Truman was not a leader who would seize the occasion and seek some sort of arrangement with the Soviet Union to prevent a nuclear arms race. Thus, rather than dealing with what Truman had actually said, which, as noted above, indicated a complete lack of understanding regarding the possibility of keeping the bomb a secret (once again, the scientific knowledge was readily available even without spies), Oppenheimer replied, "Mr. President, I feel I have blood on my hands" (ibid., p. 332).

At this point, Truman lost his patience and ended the meeting. After that, the direct line of communication between Oppenheimer and Truman was closed, and all of Oppenheimer's efforts were in vain. In the years that followed, he would fail to prevent the development of the hydrogen bomb, but chose not to resign even after Truman decided, due in no small measure to internal political pressure,[14] to accelerate the project.

As part of this process of coming to his senses, Oppenheimer realized that Rabi was right when he refused to join the project because he feared that "the culmination of three centuries of physics" would be a weapon of mass destruction. Not only did the war not advance physics, "It practically stopped it" (ibid., p. 322), and fundamentally changed its nature.[15]

If Oppenheimer was at his very best before the bomb, the picture changed very quickly, and only a few days after it was detonated, he was "a nervous wreck". It seems that the more time passed, the more aware he became of the ramifications of the project he led—not only of the hundreds of thousands of people killed, but of the new and monstrous world he had brought humanity into.

Following Japan's surrender, he drafted a final report on behalf of the Scientific Panel, in which a completely different approach emerges: "It is our firm opinion," he wrote, "that no military countermeasures will be found which will be adequately effective in preventing the delivery of atomic weapons." Moreover, not only are we "unable to outline a program that would assure to this nation for the next decades hegemony in

[14] In his memoirs, Truman repeatedly stresses that the development of the atomic bomb was entirely independent of internal politics.

[15] It is interesting to note that after Los Alamos and his hearing, Oppenheimer ceased his active involvement in physics (Schweber, 2000).

the field of atomic weapons," but "we are equally unable to insure that such hegemony, if achieved, could protect us from the most terrible destruction." Oppenheimer concluded with what appears to be an unattainable dream: "We believe that the safety of this nation ... can be based only on making future wars impossible." He himself had no idea of how that dream might be achieved. Perhaps that is why he wrote to one of his former teachers, a few days later, that he was "a stone's throw from despair". To his Harvard roommate Frederick Bernheim he wrote, "We are ... in an earnest but not too sanguine search for sanity" (pp. 313–319).

Technical Sweetness

On 7 August 1945, the day after the detonation of the bomb over Hiroshima, Oppenheimer's close friend Haakon Chevalier, sent him a note of congratulations.[16] Twenty days later, Oppenheimer replied to Chevalier that "the thing had to be done", especially "at a time when all over the world men ... *were committed as never before ... to technology as a way of life and thought*" (ibid., p. 319). Right there, Oppenheimer reveals an important aspect of the bomb for the purposes of our discussion—that it was the result of a total commitment to technology.

On the question of commitment to technology, Oppenheimer represents an approach that is reminiscent, at least to some extent, of the position taken by Heidegger. Oppenheimer says so quite clearly, "What are we to make of a civilization which has always regarded ethics as an essential part of human life, [but] *which has not been able to talk about the prospect of killing almost everybody except in prudential and game-theoretical terms?*" (ibid., p. 563). Heidegger expresses the very same idea, but in a more sophisticated way: "Technology as the highest form of *rational consciousness*, and the lack of reflection as the arranged powerlessness, opaque to

[16] It was a friendship that would cost him dearly at the time of his 1953–1954 security hearings. In 1943, when he was already director of the Manhattan project, Chevalier approached him with a proposal to spy for the Soviets. Oppenheimer decided, at a certain stage, to report the proposal to his superiors, but did not mention Chevalier (attempts to explain this in terms of a childhood experience after having tattled on someone are, in my opinion, little more than pop psychology). This lie was to Oppenheimer's detriment at the hearings. He did not deny having lied but said that he had acted like an idiot. We now know that behind Oppenheimer's dismissal and hearings stood his opposition to the development of the hydrogen bomb.

itself, to attain a relation to what is worthy of question, belong together: they are the same thing" (Heidegger, 1973, p. 99).

Both Heidegger and Oppenheimer thoroughly understood the destructive nature of technology, although, ultimately, the way in which they spoke about technology somehow ***diminished the responsibility of human beings for their own actions.*** Oppenheimer's use of the idea of "technical sweetness" (during his hearing in 1954) is a classic example of this:

> ***When you see something that is technically sweet, you go ahead and do it*** and you argue about what to do about it only after you have had your technical success. That is the way it was with the atomic bomb. (USAEC, 1954, p. 81)

This approach is based on the principle of technological determinism, which denies the responsibility of humans for their own actions. Oppenheimer, like Heidegger, and perhaps Speer, shifts blame from humans to technology—thereby absolving himself and his colleagues of responsibility for the bomb, while assuring all of us a very grim future indeed.

References

Baggott, J. (2010). *The first war of physics: The secret History of the atom bomb 1939–1949*. Pegasus Books.

Best, G. (2005). *Churchill and war*. Hambledon and London.

Bird, K., & Sherwin, M. J. (2005). *American Prometheus: The triumph and tragedy of J. Robert Oppenheimer*. Atlantic Books.

Cohen, A. (1989). *Nuclear age: A chapter in moral history (Heb.)*. Ministry of Defence.

Compton, A. H. (1956). *Atomic quest: A personal narrative*. Oxford University Press.

Daley, T. (2010). *Apocalypse never: Forging the path to a nuclear weapon-free world*. Rutgers University Press.

Greene, B. (2004). *The fabric of the cosmos: Space, time, and the texture of reality*. Alfred A. Knopf.

Heidegger, M. (1973). Overcoming metaphysics. In M. Heidegger (Ed.), *The End of Philosophy* (J. Stambaugh, Trans. pp. 84–110). Harper and Row.

Lemmerich, J. (2011). *Science and conscience: The life of James Franck* (H. A. M, Trans.). Stanford University Press.

Monk, R. (2012). *Inside the centre: The life of J. Robert Oppenheimer*. Random House.

Schweber, S. S. (2000). *In the shadow of the bomb: Bethe, Oppenheimer, and the moral responsibility of the scientist*. Princeton University Press.
Schweber, S. S. (2009). *Einstein and Oppenheimer*. Harvard University Press.
USAEC. (1954). *In the matter of J. Robert Oppenheimer: Transcript of hearing before personnel security board, Washington D.C., April 12, 1954 through May 6, 1954*. United States Government Printing Office.

CHAPTER 8

The Collapse of the Second Line of Defense: It's Not Us; It's Technology

The idea of technological determinism is not as simplistic as we would like to think. It is hard to deny that we are trapped in a technological state of being, in which freedom of choice is significantly reduced. The question thus arises whether the twentieth century can only be described through the internal dynamic of technology. Are we willing to give up that which makes us human—the ability to choose, to oppose the technological dynamic?

Technology has an internal dynamic and can, undoubtedly, change our way of thinking, but can we accept, by way of explanation of the atomic bomb, a pronouncement along the lines of "*It's the commitment to technology*"? Even if we accept technology as a state of being, as a way of thinking, pronouncements of this kind lie at the heart of generalities that are little more than superficial ideas. The idea that "the production of corpses in the gas chambers and the extermination camps" is "the same thing in its essence" as the mechanized food industry, because both are the central cause of technological being, sounds, at best, like the justification of active participants in the destruction of humanity. At worst, it sounds like superficial philosophy—in the sense that it is easier for us all to blame technology than to take responsibility for our actions. According to this approach, there is no essential difference between fertilizing the soil and building an atomic bomb. In both cases, we are challenging nature, and there is no difference between someone who sends large numbers of people their

Y. Ataria, *Genes, Technology, and Apocalypse*,
https://doi.org/10.1007/978-3-031-59197-6_8

deaths and someone who makes hamburgers at McDonald's. In any event, even if we can imagine some connection between the two, and even if Heidegger was right when he said that "the greatest care must be fostered upon the ethical bond at a time when technological man, delivered over to mass society, can be kept reliably on call only by gathering and ordering all his plans and activities in a way that corresponds to technology" (Heidegger, 2008, p. 255), putting them both in the same basket would seem to be a simplification that one would be hard-pressed to justify.

Despite the commitment to technology, the genocide of the Jews and the dropping of the atomic bombs on Hiroshima and Nagasaki cannot be reduced to mere technical actions. Indeed, a holocaust is "a catastrophe whose possibility was born of technology and cannot be prevented without technological intervention. But technology isn't enough" (Ophir, 2005, p. 590).

In the next chapter we will see that when we understand the broader context in which we act, we will also discover that the vector that points toward the beginning and the vector that points toward the end are in fact a single vector driven by apocalyptic desire. Thus, even if technology is not enough, there can be no doubt that within the framework of apocalyptic being, it is very hard to prevent the complete destruction to which it leads.

References

Heidegger, M. (2008). Letter on humanism. In M. Heidegger & D. F. Krell (Eds.), *Basic writings: From being and time (1927) to the task of thinking (1964)* (pp. 213–265). Harper Perennial.

Ophir, A. (2005). *The Order of evils: Toward an ontology of morals* (R. Mazali, & H. Carel, Trans.). Zone Books.

PART III

Apocalypse

Scholar of Kabbalah and Jewish mysticism Joseph Dan argued that apocalyptic thought is "impossible without the premise of biblical time," yet, it is equally impossible "without deviating from or even shattering this concept of time" (Dan, 2009, p. 41). Dan believed that the arrow of linear time derives from the foundational phrase: "In the beginning God created the heaven and the earth" (Gen. 1:1), which places history on an axis of time with a beginning and an end: "The concept of time presented in the Bible is linear, drawing a direct line from creation to the end of days, and it is along the continuity of this line that history unfolds" (Dan, 2000, p. 41). Apocalyptic thought develops and acts within an intellectual field in which there is "a tight bond between beginning and end" (p. 33) or, in the words of the great modernist poet T. S. Eliot, "What we call the beginning is often the end | And to make an end is to make a beginning. | The end is where we start from" (Eliot, 1972, p. 58). The logic behind this is clear: it is a Freudian-like deterministic approach (the pathology in the present necessarily attests to some previous complex). Accordingly, understanding something about the beginning also allows us to say something about the end (and vice versa), which is the very premise of those who seek the bones of the first human.

The apocalypse, as Hollywood films repeatedly demonstrate, pertains not only to the study of biblical history, but plays a fundamental role in the structure of the present, that is to say that the apocalypse is inherent to the way we think about the future of humanity. Essentially, the apocalyptic mindset or mode of being is an integral part of modernity and progress. In the words of Dan, "[All] areas of culture related to modernity, capitalism, socialism, and democracy are based on the Christian world view [particularly the book of Revelations] that inherited from Judaism [particularly the book of Daniel] the concept of time and history originating in the book of Genesis" (Dan, 2000, p. 21).

The apocalypse, according to Dan, "takes the bond between this world and the divine out of the hands of the individual, [which is why] apocalyptic literature does not preach religious observance" or call upon the individual to attain "religious perfection". Apocalyptic thought distinguishes between the righteous and the wicked, but generally as "pre-established designations, rather than descriptors assigned to individuals based on their actions or omissions". Thus, "In the apocalyptic age, devotion to religious ritual is 'too late', [because] the standing of individual human beings has already been established." The apocalypse comes to be, argues Dan, as a result of "an internal necessity stemming from the preordained nature of history ... which individuals are powerless to influence... for good or for ill" (Dan, 2009, pp. 54–61). In these circumstances, all that remains is to seek to hasten it. An individual may attempt to influence history only if "the end-time is not absolutely fixed, but depends, at least to some extent, on human beings, whose actions can hasten or delay it" (Dan, 2000, p. 56). It thus comes as no surprise that "those who believe in the ability of human beings to improve themselves and the reality around them, through the observance of God's commandments, are not stressing the apocalyptic element in their world view" (Dan, 2000, p. 38).

In the two chapters of this third part of the book, I will suggest that the atomic project and the search for the bones of the first human stem from the same motive: both symbolize the desire to be redeemed from reality and to reach the end-times. The bones of the first human have the potential not only to tell us the story of humanity's beginning and end, but also to contribute to our understanding of "human nature", allowing us to effect the necessary changes on the path to creating the "superhuman". The atomic bomb and the concentration and annihilation camps (as code names) are the means that will allow the superhuman to take over the world without difficulty.

In the first chapter, we will explore the connection between the apocalypse and the atomic bomb. We will begin by presenting Heidegger's perception of the danger posed by the development of the atomic bomb, which, in his view, lay not in annihilation itself but in changing the way in which we are thrown into the world. It will further be suggested that the bomb reveals a deep truth about the essence of technology and science: the bomb attests to the fact that Western culture is immersed in nihilism, a further step along the path of turning human beings into "standing-reserves". The bomb shows how fluid our existence has become: we, not plutonium, are the raw material of which the bomb is made. Heidegger believed that the danger is the result of technological transparency: we are thrown into the technological mode of being, yet it remains hidden from us. Not only is technology concealed, but the very fact of its concealment is forgotten, making it impossible to break free of its transparent grip. The primary characteristic of technology is its transparency; it is in the background. Due to our tendency to think of technology in terms of tools, we repeatedly fail to understand that technology is not something that can be grasped, but the field in which we act. Again and again, our focus on tools (such as the current mania surrounding AI) conceals the essence of technology.

In light of the above, we will try to decipher Heidegger's famous remark that "only a God can save us". As we will see, this is a further example of the deep connection between the bomb and the apocalyptic mode of being, between *t*echnology and apocalypse. In support of this idea, we will also turn to Stanley Kubrick's (1964) film *Dr. Strangelove or: How I Learned to Stop Worrying and Love the Bomb*, which teaches us that the connection between rationality in general and game theory in particular and the desire for total annihilation is inherent and fundamental. Rationality is not a tool by means of which release from apocalyptic being can be achieved. Rather, it acts within the apocalyptic state of mind and strives to realize the apocalyptic vision.

The second chapter examines what appears to be an inherent link between the utopian vision associated with the scientific revolution and the doctrines of the Enlightenment on the one hand, and the apocalyptic mode of being, on the other hand. Utopia, it will be argued in this chapter, cannot be thought of but in relation to apocalypse.

The scientific revolution and Enlightenment are generally believed to have released us from divinity and messianic apocalyptic vision. It is my contention that this thesis is fundamentally mistaken: the concepts may have changed, but the essence remains the same. There is a direct link between utopian thought and the apocalyptic state of mind. The apocalypse cannot be cancelled but only hastened or delayed by means of utopia. Utopia is, on the one hand, an attempt to delay the end-times, but it is also a fantasy that strives to hasten the end and rebuild the world in a way that will bring about the end of history. Utopia cannot be thought of but in relation to apocalypse. It thus follows that utopia—the good place that does not exist—cannot be thought of but in relation to dystopia—the bad place that proves, time and again, that it very much does exist.

The more we contemplate the question *What Is Enlightenment*? the more we realize that the Enlightenment project is fundamentally utopian. To a large extent, the Enlightenment project, which can also be termed a Promethean project, aims to create a new human. It seems, however, that whenever we try to realize the utopian project of creating a new human, the result is dystopic. When we examine the connection between the deepest scientific aspirations and utopian thought, we find that the two projects cannot be separated from one another. The scientific project is conducted within the framework of the apocalyptic mode of being, making it not unlikely that it will lead us to a dystopian reality. In support of this argument, we will examine the development of the idea of eugenics, from the late nineteenth century to the present, and we will see how (pseudo-) scientific discourse can be used to justify the annihilation of entire species and races and anyone considered "other" or "weak".

Although perhaps a comforting thought, such problems are not limited to dictatorships. We are now entering the age of liberal eugenics that can lead to a homogeneous and utopian society, as described in Aldous Huxley's *Brave New World*, which reveals the deep connection between enlightenment, science, rationality, and our utopian aspirations as a society. A careful reading of the book shows that science, rationality, enlightenment, and utopia are inherently linked, not only to one another, but also to the ostensible mirror image of utopia—dystopia. From this perspective, we begin to understand both the desire to discover the bones of the first human, and the desire to use all-destroying bombs: while the search for the bones of the first human allows us to understand how we might change our genetic makeup to create a perfect human in a utopian world, the atomic bomb allows us to release the next species from the sickness of this world. Both of these scientific projects occur within apocalyptic being.

I will conclude this third and final part with a return to the Archimedean point of the entire book: Kubrick's *2001: A Space Odyssey*. This film precisely defines the mode of being of this age. Kubrick essentially creates a modern myth, and since we are unable to think outside of myth, offers an accessible representation of the reality of our lives. The film seems to offer two possibilities: on the one hand, the attempt to describe the "first human" as inherently violent; on the other hand, the shifting of responsibility from humanity to technology. What we really learn from the film, however, is that the two vectors are not opposed to one another, but in fact follow the same trajectory, deriving from the apocalyptic mode of being into which we are thrown.

References

Dan, J. (2000). *Apocalypse then and now (Heb.)*. Sifre Hemed.

Dan, J. (2009). *Chaos theory and the study of history (Heb.)*. Kinneret, Zmora-Bitan, Dvir.

Eliot, T. S. (1972). *Four quartets*. Faber and Faber.

Kubrick, S. (Director). (1964). *Dr. Strangelove or: How I Learned to Stop Worrying and Love the Bomb* [Motion Picture].

CHAPTER 9

The Bomb and the Apocalypse: Only a God Can Save Us

The Danger of the Bomb

In his book *The Origin of the Work of Art*, written between 1935 and 1937 and first published in 1950, Heidegger argues that "the equipmental quality of equipment consists in its usefulness" (Heidegger, 2008a, p. 159). In other words, in order to understand what a particular piece of equipment is, we must act with it and through it in the world. It is not an object, but something through which we act in the world. It invites us to act in a certain way in the world and consequently changes the way in which we see and understand the world and thus the way in which we are thrown into the world. The knife I hold when I cut up vegetables for a salad invites me to cut differently, making the salad a different salad. A fast car invites us to take greater risks on the road, a jeep to take dirt roads.

Heidegger argues that in order to understand the essence of a particular tool, it is not enough to understand its various uses. A tool is always part of a kit, a workshop, a set of events, practices, crafts, and world views. Essentially, as we saw in Part II, technology is not just a specific set of tools but everything in the background: electricity, a system of roads, engineered vegetables, pharmaceuticals, and infinitely more. We are steeped in technology and technology is not only hidden from us, but its very hiddenness is forgotten.

Y. Ataria, *Genes, Technology, and Apocalypse*,
https://doi.org/10.1007/978-3-031-59197-6_9

From Heidegger's point of view, "Technology is ... no mere means. Technology is a way of revealing" (Heidegger, 1977, p. 12). Heidegger returns to the Greek word *poiesis* (ποίησις), from the verb *poiein* (ποιεῖν), meaning to make or create. The process of creation (craft) is thus related not only to production for a particular purpose, but also to poetics (art)—a process that gives rise to progress and to "bringing forth" that which is normally concealed. Let us begin with a simple example. A spade allows us to reveal that which was hidden beneath the soil. Of course, we can also dig with our hands, but the spade allows us to reach places that we are unable to reach with our hands alone. A pen allows us to reveal truths about our lives that we did not know until the moment that we began to write. A paintbrush allows us to reveal hidden facets of the world around us and of our own mental state. In order to paint, we also require technical knowledge—craft. An accomplished painter must also be an artisan—a necessary but by no means sufficient condition. Not everyone who knows how to hold a brush knows how to paint, which requires something more. When craft is combined with art, the result can be a revelation of truths that normally remain hidden. In this sense, the act of creating art is also an act of revealing truth.

According to Heidegger, technology, under certain conditions, allows truth to emerge. In support of this idea, he turns to the word *techne* (τέχνη): "There was a time when it was not technology alone that bore the name *technē*. Once that revealing that brings forth truth into the splendor of radiant appearing also was called *technē*" (ibid., p. 34). Heidegger thus points to the two meanings of the word. One meaning pertains to the means of producing, to the practical ability to produce (know-how), as we find in the writings of Aristotle, who recognized the fact that *techne* also belongs to *poiesis* and thus entails poetic action—a critical point, as we will see. The second meaning of *techne*, which is the more important one according to Heidegger, pertains to knowing (*episteme*) in the widest possible sense—a way of being at home in the world and understanding it. Essentially, by being thrown into the world and understanding it, we are also opening it up, and in so doing, truth can be revealed. In short, under certain conditions, *techne* reveals truth, but clearly, it can also hide the truth.

Ultimately, understanding technology as a way of knowing is necessary in order to be released from its transparent grip on us. The essence of technology has nothing to do with tools; rather, it is a way of being-in-the-world. It shapes the way in which we are thrown into the world, that is to say that it is an epistemic sense of understanding the world (note that for Heidegger, understanding is a way of being projected into the world). Only once we start to realize that technology is a mode of being, a mood, can we break free of the grip of technology. By understanding that technology neither grasps nor is grasped, but is the primary and basic way in which humans afford meaning to the world, we may begin to understand the connection between truth and release from technology.

In any event, the paradox is that as long as we remain immersed in the discourse of grasping and controlling we remain trapped in the technological state of mind. We are absolutely sure that we understand technology, yet it is merely an illusion that leads us to fall even deeper into the technological mode of being. Consequently, discourse of human control of the atomic bomb is fundamentally flawed—accuracy that hides the truth. What is really important is to change the mode of being that created the bomb in the first place, the way in which it changed the world, made it so uncanny, threatening, and distant. The great danger of the atomic bomb, Heidegger insists, lies not in total annihilation but in the loss of our ability to think, rather than merely calculate and represent (Heidegger, 1966).

Heidegger did not believe that we completely lost our ability to be in the world after the bomb, but the camouflage of technology was upgraded to the point that we are no longer capable of distinguishing between nature and technology. We drive along French country roads and admire the vineyards as if they were nature, rather than technologically engineered cultivars planted in soil worked and treated by means of a variety of technologies. It is not nature but technology, and we have lost the ability to see it. So immersed are we in the world of representations and simulacra (representations devoid of any real object, copies without origin pretending to be original) that we are convinced that what we see is the world itself. We believe that science helps us to reveal the world, but it only veils it from us. The absurd thing is (and this is the danger, according to Heidegger) that in the techno-scientific age, only that which can be represented, that is produced or treated as if it were a resource to be shipped and handled, calculated and consumed, is considered real, or, in the words of Horkheimer and Adorno, two German-Jewish thinkers of the Frankfurt School: "For enlightenment, anything which does not conform to the standard of calculability and utility must be viewed with suspicion" (Horkheimer & Adorno, 2002, p. 3).

To Heidegger, the atomic bomb and, all the more so, the hydrogen bomb, is merely the tip of the process. The catastrophe has already happened, even without the world being destroyed. The bomb tells the story of Western science and metaphysics—the metaphysics of technology: "The atomic bomb exploded long ago; namely at the moment that the human stood in insurrection against being and posited being from out of himself and made it the object of his representing. This since Descartes" (Heidegger quoted in Mitchell, 2015, p. 66; GA 15: 433).

A priori, science deals only with objects it has destroyed. Until an object has been destroyed and produced as representation, it is of no interest to science. Through this process, rather than revealing the world, scientists conceal it and render it inaccessible.[1] For Heidegger, the bomb attests to our inability to be in the world if not through technology. It is impossible to devise, plan, and carry out the atomic bomb project or the horrors of the gas chambers but in the framework of the totality of technology. Scientific application is part of the overall submission to technology. More than wiping out an immense number of people in no time, the message of the bomb is the transformation of human individuals into a mass, against their will, thereby essentially turning them into a resource—part of the production process.

The bomb also changed concepts of space and time, eliminating distance and shrinking time. If, in the concentration and annihilation camps, human beings were enclosed in a limited space in order to destroy them in a minimum of time and with maximum efficiency, in the age of the bomb, they can be destroyed in no time at all. There is no need for ghettos or trains or camps. Humans, not plutonium or hydrogen, are the raw

[1]A prominent example of this can be found in the neuroscientific study of consciousness. We expect brain-imaging research (fMRI) to tell us something about our spiritual lives. In practice, however, brain-imaging research is able to detect magnetic resonance in blood, allowing it to observe changes in blood flow between different parts of the brain. The resulting image that we see in scientific journals or popular media is of a brain marked with different colours in different areas, alongside descriptions of the particular experiences they represent: depression, schizophrenia, etc. In reality, however, it is the result of a statistical production process that transforms information received regarding blood flow into a coloured image. When we look at it, we get the feeling that we are looking into our very soul, when what we are actually seeing is a neuroscientist's interpretation presented as fact: a perfect simulacrum. The problem is not only the illusion that we are watching the soul in action, but the fact that modern neuroscience knows absolutely nothing about the subjective human experience. Ever greater funds are channelled to brain research and it only pushes us further away from understanding the subjective experience. What neuroscience is showing us, in effect, is how various technologies allow us to understand natural processes, but—and this is the cardinal point—what they offer is not neutral observation, but a process of representation and production. This process does not reveal truth, but merely conceals it further—all within the framework of the illusion that we are getting closer to truth.

material of the bomb, and we are in a constant state of readiness for annihilation—as a result of which, our existence becomes fluid.

In the glaring light of the explosion, there is no place to hide, and the absent-present dialectic from which truth emerges is also denied, meaning that truth can no longer be revealed. If the technological takeover of our lives was somewhat hidden before the development of the bomb, along came the bomb and revealed that we are all a standing-reserve, ready for annihilation, our existence liquidated. That is why, according to Heidegger, the bomb is destructive even without destroying. The bomb changed the reality of our lives, even without exploding. It revealed the world as it has been since long before the bomb: a technological desert. It is precisely at this low point, however, that the possibility of change emerges: "Where the danger is as the danger, there the saving power is already thriving also. The latter does not appear incidentally. The saving power is not secondary to the danger. The selfsame danger is, when it is as the danger, the saving power" (Heidegger, 1977, p. 42). As we try to understand this idea, we will undoubtedly return to some of the ideas discussed earlier.

As we have seen, for Heidegger, the biggest problem with technology is our inability to understand that it is not a tool but an all-encompassing mode of being—a particular way of being thrown into the world. In other words, the Modern Era is characterized by our total falling into technological being—a state in which everything is a standing-reserve, including human beings.[2] It is a way of thinking, yet it remains transparent. The more we engage with the dangers of smartphones or self-driving cars or Facebook addiction or ChatGPT, the more we fail to recognize the true essence of technology. We have seen that the sense of loss of control is, for Heidegger, the first step to understanding the essence of technology. The mistake would be to think that we must try to grasp it and control it, because that is exactly when we fall even deeper into technological being. In order to achieve a free relationship with technology, rather than opposing it, we must, in a certain sense, wait, step back, and allow it to exhaust itself. As long as we are in an inauthentic state of being, and are part of the mass (the They—*das Man*), constantly falling further and further into technology, our ability to understand the state we are in is diminished. In

[2] The theological connotation of the concept of "falling" is not a coincidence and should not be ignored—here or elsewhere in Heidegger's writings. For an extensive discussion, see (Burdett, 2014). Indeed, it seems that Heidegger's strict religious upbringing deeply influenced his philosophy.

other words, our ability to take part in a process of revealing truth dissipates, and with it, the possibility of existing freely alongside technology. Heidegger insists, however, that it is precisely where the danger is greatest that salvation lies. The atomic bomb is a scientific and technological triumph and, at the same time, the greatest tragedy we can possibly imagine. The connection between the two is not a coincidence. The bomb allows every person to understand the essence of technology—an autonomous system that acts from within itself, detached from humankind, and transforms everything into a resource. It is for this very reason that this moment, precisely because of its horror, can be a lightning-flash of truth. The bomb can allow us to wake up, not only as individuals, but as a society. The atomic bomb is not an accident, an exception that proves the rule. It is the culmination of a techno-scientific project, the moment at which the dynamic that was always in the background suddenly becomes something that can be grasped, gripped, if only for a moment, if only to be immediately let go of, since technology is not something that can be grasped.

The Bomb As Techno-Scientific Poetics

Heidegger believed that art, although created with tools, can help us understand the essence of technology: "The equipmentality of equipment first expressly comes to the fore through the work [of art] and only in the work ... In the work of art the truth of beings has set itself to work" (Heidegger, 2008a, pp. 161–162). Truth, according to Heidegger, is not the identity of the representation of a thing and the thing itself, or between a statement and the content to which it refers. When Heidegger speaks about truth "setting itself to work", he is referring to the act of unconcealing, essentially asking: "What makes a true thing true?" In Western metaphysics, truth is defined as the identity of signifier and signified, the representation and the object itself, "the adequation of thing to intellect" (Heidegger, 2008b, p. 118). Heidegger believed that this definition misses the essence of truth, reduces the idea of truth.

Heidegger sought to go beyond the idea of representation, and in order to do so, returned to the word "truth" in Greek, *aletheia* (ἀλήθεια), meaning unconcealment. According to this approach, truth and untruth must be thought of in terms of the unconcealment (disclosedness) and concealment (undisclosedness) of being, respectively. It is important to note that truth and untruth in this context are not things or representations of things, but interdependent processes: "Untruth must derive from the essence of truth. ... Only because truth and untruth ... belong together" (Heidegger, 2008b, p. 128). The disclosure of truth is a process in which truth and untruth, as two essential forces, strive with one another—a process in which concealment and unconcealment are in dialogue, in an interplay of light and darkness, absence and presence. Art brings forth (a process of truth production) that which was previously concealed. The activity of the sculptor, inasmuch as it is a process of revelation, offers a helpful illustration of this idea: "To be sure, the sculptor uses stone just as the mason uses it, in his own way. But he does not use it up. That happens in a certain way only where the work miscarries" (Heidegger, 2008a, p. 173). As the sculpture is revealed, the material is not used up, but truth is produced and brought forth. Artistic creation is "the bringing forth of the unconcealment of beings" (p. 196).

According to Heidegger, the work of art "moves the earth itself into the open region of a world and keeps it there. *The work lets the earth be an earth* [emphasis in the original]" (p. 172). For Heidegger, inspired by Heraclitus, the earth is that which lies "beyond the realm of meaning", representing the impenetrable and the irrational, the "impossibility of revealing all aspects of beings", what humans cannot control: "When we seek to give it form, the earth withdraws; when we seek to acquire it, it eludes us". The earth, in its essence, is "self-secluding (*Sichverschliessende*)" (Mansbach, 1997, pp. 84–85). In the context of techno-science, the earth can be said to cause "every merely calculating importunity upon it to turn into a destruction" (Heidegger, 2008a, p. 172). In other words, the earth is precisely what is beyond representation and, as noted, science, according to Heidegger, can only act on that which it itself produces and represents. In this sense, the earth marks the limits of science: that which does not submit to calculative thinking.

In the techno-scientific age, the following question arises: is there any conceivable action that would let the earth be an earth more than the atomic bomb. The bomb is more like sculpture and revelation than using matter up. The bomb is sculpture on a new scale: techno-scientific poetics. The bomb as poetics, as a lightning-flash of truth, lets the earth come forth from itself and for itself, as it is. In the bomb, which links danger and salvation, lies the potential to reveal the earth, once again, outside of technological metaphysics—as something more than a source of energy. In this sense, Hiroshima is identical to the Greek temple (Fig. 9.1):

Fig. 9.1 Hiroshima as a Techno-Scientific Temple

> A building, a Greek temple, portrays nothing. It simply stands there in the middle of the rock-cleft valley. The building encloses the figure of the god, and in this concealment lets it stand out into the holy precinct through the open portico. By means of the temple, the god is present in the temple. This presence of the god is in itself the extension and delimitation of the precinct as a holy precinct. The temple and its precinct, however, do not fade away into the indefinite. It is the temple-work that first fits together and at the same time gathers around itself the unity of those paths and relations in which birth and death, disaster and blessing, victory and disgrace, endurance and decline acquire the shape of destiny for human being. (Heidegger, 2008a, p. 167)

Of course the atomic bomb also presents dangers (not in the obvious sense of killing large numbers of people, but in the sense of turning people into standing-reserves), but it is, at the same time, a work of art that discloses truth. I will elaborate on this point.

Much has been written about the relationship between craft and art. It is generally claimed that not every artisan is also an artist, but every artist is also an artisan. Heidegger believed that such arguments miss the essential point:

> However usual and convincing the reference may be to the Greek practice of naming craft and art by the same name, *technē*, it nevertheless remains oblique and superficial; for *technē* signifies neither craft nor art ... The word *technē* denotes rather a mode of knowing ... a bringing forth of ... what is present as such out of concealment ... into the unconcealment of its appearance.... The artist is a *technitēs* not because he is also a craftsman, but because both the setting forth of works and the setting forth of equipment occur in a bringing forth that causes beings in the first place to come forward and be present in assuming an outward aspect. (Heidegger, 2008a, p. 184)

The concept of *techne* seems to combine praxis with art, the technical with the true. If we look at the bombing of Hiroshima as *the* act of artistic creation, one that exposes the essence of all technology and not only the "equipmental quality" of this or that piece of equipment, then Hiroshima is the truth that stands at the very heart of being: "the clearing center [that] encircles all that is" (ibid., p. 178).

Heidegger believed that "science is not an original happening of truth, but always the cultivation of a domain of truth already opened" (ibid., p. 187). If so, the scientists in Los Alamos did not invent anything, but merely engaged in engineering—the shaping of reality according to the representation and production at the heart of the techno-scientific mindset. Put another way, the atomic scientists set about domesticating the stubborn earth. And is there a better way to flatten any field and prepare it for redesigning than an atomic bomb? In this sense, the atomic bomb is an architectural project par excellence, and the only architectural project possible in the technological age.

In any event, the atomic bomb reveals a more complex truth: ultimately, technology brings to a head the struggle between the domestication of the earth and its return at humanity's expense. To understand this idea better, we must return to two fundamental principles in Heideggerian thought: the essence of truth and the strife between world and earth, and the connection between them. According to Heidegger, "The essence of truth is un-truth", but that is not to say, he stresses, that "truth is at bottom falsehood", that "truth is never itself but, viewed dialectically, is also its opposite". The essence of truth is thus "that opposition ... which subsists between clearing and concealing. ... The essence of truth is, in itself, the primal strife" (p. 180).

Let us now turn to the strife of world and earth:

> The earth is not simply the closed region but rather that which rises up as self-closing. World and earth are always intrinsically and essentially in conflict, belligerent by nature. Only as such do they enter into the strife of clearing and concealing. Earth juts through the world and world grounds itself on the earth only so far as truth happens as the primal strife between clearing and concealing. (p. 180)

The atomic bomb, as the greatest artwork of the techno-scientific age, and perhaps the only possible artwork in the ultra-technological age—that is, within the technological ghetto—presents deep and irreconcilable strife between the domestication of earth and its annihilation. In this context, the search for the bones of the first human is an attempt to exhume the remains of previous versions of the struggle between living world and earth, in order to understand the nature of this struggle better.

The technology of the atom allowed the earth return to what it always was. In the process, however, it left no room for humans on the face of that earth. Humans seek to domesticate the earth, to push it out of the world. Somewhat surprisingly, the atomic bomb, which represents a certain high point of our ability to control our surroundings and, as such, the climax of the techno-scientific programme, actually marks the return of the earth, at our expense. The atomic bomb allows the earth to emerge once again,

> The final chapter of this consummation will consist in the earth itself blowing up and current humanity disappearing. That will not be a misfortune but, instead, the first purification of being [*Reinigung des Seins*] from its most profound deformation on account of the supremacy of beings [*Seiendes*]. (Heidegger, 2017, p. 187)

The renewed emergence of the earth after the bomb, the renewed emergence of that which does not submit to calculative thinking, that which refuses to be represented, refuses to be disciplined, can bring about a number of different results, one of them being that the earth will swallow humankind. It is also possible, however, that the earth's breaking out will allow renewed existence—faithful to the poetics of the earth—existence that, like large works of art, has brighter colors. In this new framework, the earth will rise up and it may be possible to begin anew—life in which it is possible to think rather than calculate, life that is not merely technical, life in which care (a central concept in Heidegger's early philosophy) replaces the desire to dominate (the basis of the techno-scientific revolution). There is also the possibility, which seems more likely from the perspective of 70 years, that the bomb is, in fact, a giant step into nihilism—another milestone in the West's fall into the void, a further step towards turning human beings into standing-reserves. If this is indeed the case, then only divine intervention can save us.

Only a God Can Save Us

In Heidegger's well-known interview to *Der Spiegel*, conducted in 1966 and published (in keeping with the conditions he stipulated) only after his death, he famously asserted that, "only a god can save us."

> Philosophy will be unable to effect any immediate change in the current state of the world. This is true not only of philosophy but of all purely human reflection and endeavor. Only a god can save us. The only possibility available to us is that by thinking and poetizing we prepare a readiness for the appearance of a god, or for the absence of a god in [our] decline, insofar as in view of the absent god we are in a state of decline. (Heidegger, 1981, p. 57)

The expression "only a god can save us" appears to have been said in the context of the essence of technology, that is, the "enframing" (*Ge-Stell*)—a concept that represents complete absorption into the totality of technology, that is, life under the principle of challenging. In essence, within the domain of technology (enframing), nothing remains unchallenged. When Heidegger speaks of modern technology, he is, in effect, speaking about life in the Modern Era, within an overarching framework: everything, humans included, is a standing-reserve. No land is truly beyond the control of technology. It is defined for construction, industry, military training, agriculture, or infrastructure. Even the areas in which we hike are defined as nature reserves. This is not only true of land, but of a broad range of materials: science is in a constant race to learn how to use various materials for a variety of purposes—all in the interest of making processes more efficient and leaving nothing unexploited. But even if we were to cite an infinite number of examples, we would still not understand what it means to live within the totality of technology. Living according to the enframing principle means that we no longer have the ability to grasp our surroundings and ourselves but in the framework of challenging. The connection between us and our surroundings, between us and the world, is not a connection of care or involvement, or of intimacy, but of domination, representation, and productivity—and we are unable to see our surroundings in any other way. We think of ourselves in the very same terms—always trying to figure out how to use our time more efficiently. We go on holiday, but that has nothing to do with actual freedom. Holidays also exist within the framework of total efficiency—when, where, and how much. This is the enframing within which we are trapped; it is the modern condition, life in the technological age. Everything functions based on the principle of efficiency. Everything functions, to the point that we no longer have the ability to exist on the face of the earth:

> Everything is functioning. That is precisely what is awesome [*Unheimliche*, uncanny], that everything functions, that the functioning propels everything more and more toward further functioning, and that technicity increasingly dislodges man and uproots him from the earth [*Erde*]. I don't know if you were shocked, but [certainly] I was shocked when a short time ago I saw the pictures of the earth [*Erde*] taken from the moon. We do not need atomic bombs at all [to uproot us]—the uprooting of man is already here. All our relationships have become merely technical ones. It is no longer upon an earth that man lives today. (Heidegger, 1981, p. 56)

The fatal mistake is to think that limiting screen time, for example, brings us closer to living freely alongside technology. But what solution does Heidegger offer? It seems that over the years, Heidegger became increasingly pessimistic about the ability of humans, who already perceive themselves within the framework of technological enframing, to live freely alongside technology. Indeed, if "only a god can save us", it is unclear what role there is left for humans to play in bringing about their release from technology.

To return to the *Der Spiegel* interview, research in recent years has shown that the interview was very well prepared. Heidegger knew the questions in advance and carefully prepared his answers. The interview can be considered a kind of testament on Heidegger's part—both a statement of defense for his actions and a summary and popularization of his philosophy. The assertion "only a god can save us" thus assumes great significance. It seems that Heidegger, at this point in his life, did not believe that humankind could change its condition. This is not entirely accurate, since Heidegger did believe that engaging in poetry and adoptive meditative thinking prepared the ground for the arrival of new gods.

The statement "only a god can save us" should be read in the context of another statement of Heidegger's in the same interview: "We do not need atomic bombs at all [to uproot us]—the uprooting of man is already here" (ibid., p. 56). From Heidegger's perspective, the development of the atomic bomb is testimony to a techno-scientific poetics that rendered the world a representation—a world in which we have lost the ability to be, to feel at-home within it. His appeal to a god shows that the roots of his thought were firmly planted in the eschatological mode of being and, as Nancy (2017) demonstrates, his thought is largely an attempt to constitute another beginning, with a new ending in mind.[3] It is important to stress, here, that it would be a fatal mistake to relate to Heidegger as having misunderstood Western philosophy. It is in fact quite possible that Heidegger absolutely understood the foundations of Western philosophy—leading perhaps to the disturbing conclusion that the connection between the foundations of Western philosophy and anti-Semitism is not coincidental.

It is unclear whether there is a connection between the god that Heidegger refers to in the 1966 interview and the gods that feature in his later thought, which may refer to the Greek gods, in light of the influence exerted on him by his favorite poet, Hölderlin. In any event, his later eschatology—in contrast to his 1927 book *Being and Time*—appears to relate less to the personal and more to humanity as a whole.

As we have seen, Heidegger believed that we are trapped in the metaphysics of technology. For Heidegger, the atomic bomb and, to a lesser extent, the Nazi movement were evidence of the inability of philosophy—or any entirely human endeavor, for that matter—to emancipate humanity. Hence his expectation from the gods: "the time of the gods that have fled and the god that is coming" (Heidegger, 1949, p. 313). The plain meaning of this phrase is that we are thrown into history at a moment in

[3] Some have argued that it would be a mistake to view Heidegger's position in the *Der Spiegel* interview as a a dramatic change in his thought. One of the central concepts in his book *Being and Time* is "Being-towards-death": "This certainty, that 'I myself am in that I will die,' is *the basic certainty of Dasein itself.* ... Insofar as I am [*sum*], I am *moribundus* [in dying]. *The* MORIBUNDUS *first gives the* SUM *its sense. ... Death is the utmost, though indefinite, yet certain possibility* in which Dasein stands before itself" (Heidegger, 1985, pp. 316–317) (emphasis in the original). Based on Heidegger's lectures from 1920-1921 (Heidegger, 2004), Burdett (2014) argues that Being-towards-death derives from religious experience—early Christian eschatology (Heidegger was influenced by Albert Schweizer). According to this approach, *Being and Time* is the result of Heidegger's secularization of these early ideas.

which the gods have fled, for they have no place in technological being, and all we can do is wait for the god to come and redeem us.

At this point, it is interesting to return to Jacques Ellul, author of *The Technological Society* (1964)—one of the most important elaborations of Heidegger's approach to the modern condition in the context of technology, as presented in his 1954 *The Question Concerning Technology*. As we have seen, Ellul's is perhaps the most extreme voice regarding the idea of technology as an autonomous and deterministic system. Ellul's point of departure is the expectation of the coming of the gods—as the only possibility for salvation, in the sense of freedom from technology.

Although Ellul does, at times, express reservations about his own position, as presented throughout his writings, it is hard to see, from his or Heidegger's perspective, how humanity might develop a free relationship with technology. According to Ellul the technological system lies somewhere between determinism and fatalism. The possibilities for freedom from technology, presented by Ellul, are generally remote and do not depend on human action, or, to be more precise, depend on an apocalyptic approach.

The apocalypse, according to Ellul—contrary to the generally accepted view—does not pertain to the future and is not beyond history. Rather, it is divine revelation within present history, here and now. Since humankind has fallen into an autonomous and deterministic technological mode of being, only an apocalypse can save us.

Having said that, let us now return to Heidegger. As we have seen, for Heidegger, truth is a happening or event that, through human action, brings forth the unconcealment of being. Heidegger goes back to the literal, original meaning of the Greek word for truth, *aletheia* (ἀλήθεια), which is unconcealment. The word apocalypse, in Greek (*apokalupsis*, ἀποκάλυψις), also means uncovering or unveiling, signifying true revelation.[4] I would thus suggest that the concept of "apocalypse" represents not only the idea of the end of history and the end of historical time, but is inherently related to attempts to reveal absolute truth—with the two vectors (final destination, revelation of truth) constituting two sides of the same coin.

[4] The adjective ἀληθής (*alethes*) is defined as follows in the Liddell-Scott-Jones (LSJ) Greek-English lexicon: "unconcealed, so true, real, as opposed to false, or to apparent". The lexicon notes that *alethes* derives from *a-* (negation) and the verb *letho* (λήθω) or *lanthano* (λανθάνω), "to escape notice, be unobserved, remain unknown or hidden". The primary connotation of *aletheia* is thus something unconcealed from view or awareness. The verb *apokalupto* (ἀποκαλύπτω) signifies uncovering something that was previously hidden (disclose, reveal, unmask)—from *kalupto* (καλύπτω), to cover or veil, and the prefix *apo-*, denoting "away from" or negation. The connotation here is one of active uncovering, unveiling, or even denuding.

Artwork is a process in which truth is revealed, a process of unconcealment of truth. So too, apocalypse is an act of uncovering of truth. In art, we pull aside the curtain concealing the essence of the thing, in a process of "disclosure of the particular being in its Being" (Heidegger, 2008a, p. 164). Apocalypse, on the other hand, is a process in which we expose fundamental truths, as if with bleeding fingers: while tools may be used to produce art, the apocalypse is completely unmediated, putting us in direct contact with the truth. In both cases, the process teaches us something about the truth, but apocalypse is the more active of the two.

The truth, in the case of the atomic bomb, is the truth of techno-science: while techno-science purports to allow us to master nature and adapt the world to the needs of *Homo sapiens*, it in fact transforms the earth into a place in which we can no longer exist (unworld): "Technicity increasingly dislodges man and uproots him from the earth. ... It is no longer upon an earth that man lives today" (Heidegger, 1981, p. 56).

For Heidegger, the essence of technology is intrinsically linked to apocalyptic being—geared towards death and driven from within itself. The crucial element that emerges here is the tendency of technology to destroy itself—in anticipation of the arrival of the new gods. It thus appears to follow that: (1) if the source of the word nihilism is *nihil*, nothing, and (2) if technological metaphysics is the culmination of nihilism, and (3) if the atomic bomb is the culmination of that process, then the essence of technology, for Heidegger, is annihilation—that is to say that technology essentially aspires to annihilate reality and, in this sense, Hiroshima was just the beginning.[5]

How I Learned to Stop Worrying and Love the Bomb

Stanley Kubrick's (1964) film *Dr. Strangelove or: How I Learned to Stop Worrying and Love the Bomb* teaches us not only how easy it is in the age of the red button to bring about the end of the world, but also, that we

[5] One of Heidegger's most prominent and original students was Hans Jonas. In his seminal work, *The Imperative of Responsibility*—which was greatly influenced by Heidegger, although Jonas was never naïve about his teacher's true nature—he argues that it is actually the slower act of self-destruction, that is, the destruction of the ecosystem, that is the more palpable disaster. For the purposes of the present discussion, there is no need to determine whether Jonas was right. Like Heidegger, Jonas acted and thought within an apocalyptic perception of reality, and he too—once again, just like Heidegger—linked the rise of techno-science to the apocalypse. Thus, from a metaphysical perspective, there is no real distance between Jonas and Heidegger. The only question is what will bring about the end.

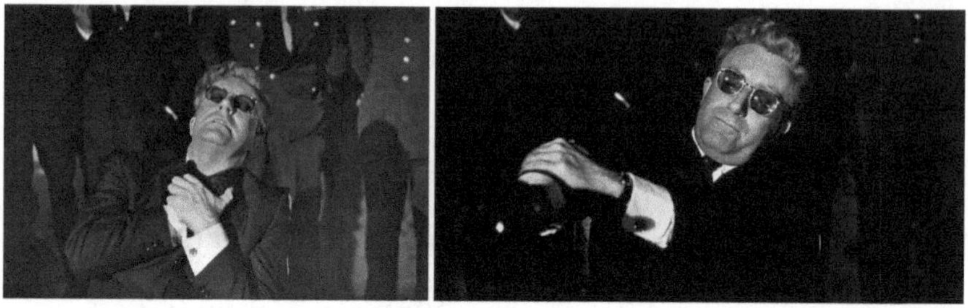

Fig. 9.2 Dr. Strangelove: The Desire for Self-Destruction Is Inherent to Rational Thought

long for that end—as the biblical Isaac longed to be bound. In other words, Kubrick exposes the apocalyptic mode in which we all exist, in which purely rational discourse conceals a deep-seated desire for self-destruction.

The main roles in the film are played by British actor Peter Sellers. He is Dr. Strangelove, Captain Mandrake, and President Muffley (and was originally cast as Air Force General Ripper as well). Dr. Strangelove is a scientist, a Nazi, and scientific adviser to the president of the United States. He suffers from anarchic hand syndrome (a condition in which a person loses control over their hand and feels as if it is controlled by some external force), at times performing a Nazi salute, at times strangling himself (Fig. 9.2).

On a superficial level, Dr. Strangelove is the president's figurative right hand, but his own literal right hand is loyal to Hitler and the Nazi regime. For example, when a miracle occurs and he stands up from his wheelchair, he cries, "Mein Führer, I can walk!" (This statement reminds us of Heidegger's words: "The Führer himself and he alone is German reality and its law.") His attempts at self-strangulation might thus be explained as attacks by his right hand, loyal to the Nazi regime, for his treasonous collaboration with the Americans. It is interesting to note that in some cases of anarchic hand syndrome, the affected hand may seek to satisfy repressed desires (inappropriate touching, for example).

On a deeper level, however, Dr. Strangelove is the president's own right hand, representing his concealed and unconscious desire for Nazi-like solutions. The message is that in every regime, there are hidden desires for unlimited power and absolute solutions. The strangling hand reminds us that absolute, scientific, and rational solutions are driven by hidden desires for self-destruction, which lie at the heart of modern society—at the heart of the Enlightenment mode of being.

From an engineering perspective, the Nazi solution constituted an innovation, both as an act and a way of thinking, although, on a deeper level, it stemmed from a messianic-apocalyptic vision of the end of the world—an approach that historian Saul Friedländer (1997) called "redemptive anti-Semitism" and hinted at in the very term "final solution": "In this type of anti-Semitism, the struggle against the Jews acquires the aura of a religious, apocalyptic war; the removal of the Jews and their alleged influence becomes a lever for cosmic deliverance" (Fraenkel, 1999). In his influential novel, *The Man Without Qualities*, the Austrian writer Robert Musil describes such a time as "full of expectation, impatient, wayward, and miserable, the messiah for whom it was hoping and waiting was not yet in sight" (Musil, 1996, p. 437). These visions also lie at the heart of strategic models taught at universities as holy writ.

Game theory, for example, which began to dominate economic, political, and strategic discourse during the Cold War, is driven by the approach developed by the Hungarian-American mathematician John von Neumann, as a consultant for the RAND Corporation. On a fundamental level, von Neumann's theory is based on the same messianic-apocalyptic principles that drove Hitler: all-or-nothing and zero-sum games, in which one destroys the other or is destroyed. It thus comes as no surprise that when Hitler understood that the war was lost, he issued the Nero Decree, ordering the destruction of the Reich.

Dr. Strangelove teaches us that this very apocalyptic mode of being and thinking sits at the heart of enlightened, democratic society—at the heart of scientific being, if you will. There is no way of thinking of the atom but in terms of total annihilation. Asking whether an atomic war should be waged, argues Israeli military strategist Yehoshafat Harkaby, is like asking what is worth committing suicide for (Harkaby, 1990). As noted, in the atomic age, deterrence is "not due to the certainty of the threat being carried out, but due to the uncertainty of it not being carried out. It is thus not the certainty but the doubt that deters. … In a conflict between the powers, the deterring side does not threaten that its opponent will be destroyed or 'die', but that both of them will 'die', essentially threatening mutual suicide" (pp. 221–222). Dr. Strangelove and General Jack Ripper are two sides of the same coin. It is no coincidence that both characters' names evoke sexual deviance: Strangelove, obviously, but also Jack D. Ripper—a reference to the Whitechapel Murderer, who targeted prostitutes. General Ripper, a US Air Force commander, believes that the Russians have contaminated the drinking water in Florida and decides to launch a nuclear war that can only be stopped with the secret code POE (peace on Earth), known only to him. Note that the code is not (only) a joke or a word game, but a deep truth in the apocalyptic-messianic-nuclear age: the atomic bomb is a means to achieving world peace (the very claim made by some of the atomic scientists).

It is hard to imagine a more estimable figure in the eyes of John von Neumann, who supported the idea of a first-strike attack, than the general who went insane by acting in an entirely rational fashion. Ripper is the anarchic right hand, obeying a different voice, representing absolute rationality, whereby a first strike is essential for victory. Yet, as we have seen, the hand that obeys the voice of rationality is driven by an apocalyptic desire—as Dr. Strangelove oscillates between "Heil Hitler!" and self-strangulation—which are, in fact, two sides of the same coin. The same is true of General Ripper, who ultimately acts in such a way as to bring about his own death—killing himself in the knowledge that his death will lead to the end of the world.

Taken together, Kubrick and Heidegger teach us that in order to understand where we are headed, we must think of humankind in apocalyptic-technological terms. It is an inevitable (albeit not insurmountable) dynamic, rooted in Western civilization's most basic collective unconscious. Thus, even if the world is not destroyed by a nuclear accident, the apocalypse is inevitable because the atomic bomb is in the background of everything we do, the context within which we operate, the world into which we are thrown.[6] In this sense, Heidegger was right: technology is the mode of being into which we are thrown, and it is apocalyptic, a train with one final stop and everyone knows what it is.

The atomic bomb is the cultural, technological, and scientific culmination of civilization and, at the same time, the realization of our collective unconscious—burning apocalyptic desire. And let us not forget that the apocalypse is "the aim towards which all previous empires developed for centuries, if not longer" (Dan, 2000, p. 28). Therefore, we must not, even

[6] To date, more than 550 accidents have occurred at nuclear reactors in the United States alone. To illustrate just how great the danger is, in 1995, a Russian officer mistook a Norwegian weather rocket for a ballistic missile launched from an American nuclear submarine and Boris Yeltsin became the first Russian president to have the "nuclear suitcase" open in front of him. Fortunately, Yeltsin realized that it was a mistake (Cirincione, 2008, p. 382).

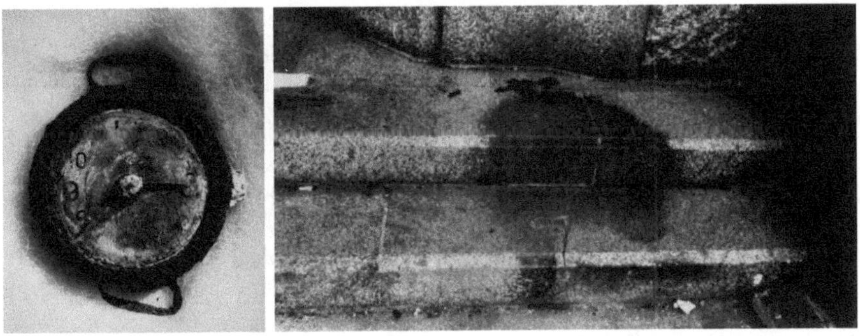

Fig. 9.3 The point Zero of Western Culture
The atomic bomb redefined the axis of time, affording the end unbearable over-presence in the here and now; no longer Heidegger's Being-towards-death, but a presence imbued with death-that-already-was. The apocalypse is part of the present moment—not merely as an unconscious religious-messianic possibility, but as part of the frightfully banal and rational cultural/entertainment present Left: A watch from the ruins of Hiroshima Right; A human shadow, Hiroshima

for a second, think of the bomb as the nadir of the human race. On the contrary, the bomb succeeded in unifying the desire for beginnings and ends. The bomb is the end and the beginning fused into a single moment. Start and finish met at Nagasaki and Hiroshima.

The atomic bomb is, as we have seen, the result of the apocalyptic mode of being. No less importantly, however, it is a product of the vision of Enlightenment, or, more accurately, the vision of Enlightenment itself developed within the framework of apocalyptic being (the Enlightenment envisaged the creation of a new human, perhaps a superhuman—beneath the desire to create a new human stands the desire to destroy the "old" human). It is, therefore, almost inevitable, or even natural, that the culmination of the Enlightenment would be a world-destroying bomb (Fig. 9.3).

References

Burdett, M. S. (2014). *Eschatology and the technological future*. Routledge.
Cirincione, J. (2008). The continuing threat of nuclear war. In N. Bostrum & M. M. Ćirković (Eds.). Oxford University Press.
Dan, J. (2000). *Apocalypse then and now (Heb.)*. Sifre Hemed.
Ellul, J. (1964). *The technological society* (J. Wilkinson, Trans.). Vintage Books.
Fraenkel, D. (1999). The abnormality of the normal. *Yad Vashem Studies, 27*, 429–439.

Friedländer, S. (1997, 2007). *Nazi Germany and the Jews*. HarperCollins.
Harkaby, Y. (1990). *War and strategy (Hebrew)*. Ma'arachot.
Heidegger, M. (1949). Hölderlin and the essence of poetry. In M. Heidegger (Ed.), *Existence and being* (pp. 291–315). Vision.
Heidegger, M. (1966). *Discourse on thinking* (J. M. Anderson, & E. H. Freund, Trans.). Harper and Row.
Heidegger, M. (1977). *The question concerning technology and other essays* (W. Lovitt, Trans.). Harper & Row.
Heidegger, M. (1981). "only a god can save us": The Spiegel interview (1966). In T. Sheehan (Ed.), *Heidegger: The Man and the Thinker* (W. J. Richardson, Trans., pp. 45–67). Precedent Publishing.
Heidegger, M. (1985). *History of the concept of time: Prolegomena*. (T. Kisiel, Trans.). Indiana University Press.
Heidegger, M. (2004). *The phenomenology of religious life*. (M. Fritsch, & J. A. Gosetti-Ferencei, Trans.). Indiana University Press.
Heidegger, M. (2008a). The origin of the work of art. In M. Heidegger & D. F. Krell (Eds.), *Basic writings: From being and time (1927) to the task of thinking (1964)* (pp. 139–212). Harper Perennial.
Heidegger, M. (2008b). On the essence of truth. In M. Heidegger & D. F. Krell (Eds.), *Basic writings: From being and time (1927) to the task of thinking (1964)* (pp. 111–138). Harper Perenniel.
Heidegger, M. (2017). *Ponderings XII–XV: Black notebooks 1939–1941* (R. Rojcewicz, Trans.). Indiana University Press.
Horkheimer, M., & Adorno, T. W. (2002). In G. S. Noerr (Ed.), *Dialectic of Enlightenment* (E. Jephcott, Trans.). Stanford University Press.
Kubrick, S. (Director). (1964). *Dr. Strangelove or: How I Learned to Stop Worrying and Love the Bomb* [Motion Picture].
Mansbach, A. (1997). Overcoming anthropocentrism: Heidegger on the heroic role of the works of art. *Ratio, 10*(2), 157–168.
Mitchell, A. J. (2015). *The fourfold: Reading the late Heidegger*. Northwestern University Press.
Musil, R. (1996). *The man without qualities: A sort of introduction and pseudoreality prevails* (S. Wilkins, Trans.). Vintage International.
Nancy, J. L. (2017). *The banality of heidegger* (J. Fort, Trans.). Fordham University Press.

CHAPTER 10

Utopia and Dystopia

Utopia: An Introduction

The concept of utopia derives from a work published by the English jurist Thomas More in 1516, the Latin title of which translates "On the Best State of a Republic and on the New Island of Utopia". Any attempt to define the concept is undoubtedly influenced by the term itself, which is a pun on the Greek words *eutopia* (εὐτόπια = εὐ + τόπια), "good-place", and *outopia* (οὐτοπία = οὐ+τοπία), "no-place". Utopia is thus the good place that does not exist. Although the concept of utopia was conceived by More, the idea of a perfect state can already be found in Plato's *Republic*. Indeed, as historical research has shown, utopian writing has been a part of philosophical thought from the very beginning.

There appears to be a direct link between utopian thought and apocalyptic being. As we have seen, the apocalypse cannot be canceled but only hastened or delayed, and utopia is the means by which this can be accomplished—albeit within an inherent paradox. On the one hand, utopia is an attempt to delay the end of days or, in the words of Joseph Dan, "'Delaying the end' is … not effected directly, but by planting the tender shoots of future reality within the currently flawed state of the world." On the other hand, utopian thought also includes the fantasy of hastening the end: "Within every utopian act lies the hope that the wonderful model devised by the utopianists will so impress the powers of history that it will hasten

Y. Ataria, *Genes, Technology, and Apocalypse*,
https://doi.org/10.1007/978-3-031-59197-6_10

the universal end". Utopia may thus be considered "an intrusion of the apocalyptic vision into history on a small and gradual scale". At the same time, however, we cannot ignore the fact that the utopian act also entails "the aspiration to change the face of the world and rebuild it in a new format, thereby precipitating the end of history" (Dan, 2009, p. 57).

Scholar of Kabbalah Gershom Scholem saw the connection between messianism, apocalypse, and salvation as stemming from opposing vectors: on the one hand, the forces of innovation, which actually look toward the past "to the return and recreation of a past condition ... Here hope is turned backwards to the re-establishment of an original state of things and to a 'life with the ancestors'"; on the other hand, the utopian forces that push forward and are "nourished by a vision of the future and receive utopian inspiration. They aim at a state of things which has never yet existed." There is a tension between the two vectors, since "even the restorative force has a utopian factor, and in utopianism restorative factors are at work". We can therefore say that every utopia imagines a future that also flows from "the restoration of what is ancient, bringing back that which had been lost; the ideal content of the past at the same time delivers the basis for the vision of the future ... The completely new order has elements of the completely old" (Scholem, 1995, pp. 3–4).

Utopia is related to time and place; utopian thinking requires an imagined geographical space in which to establish the ideal state and society. More's book was published some 25 years after the "discovery" of America, and that event appears to have afforded Europeans the opportunity to imagine a new, utopian space. European geography was thus perceived as an impediment to religious renewal and transformation. It is interesting to note, in this context, the words of Austrian-Jewish philosopher and educator Martin Buber: "The vision of rightness in Revelation is realized in the picture of a perfect time—as messianic eschatology; the vision of rightness in the Ideal is realized in the picture of *a perfect space*—as Utopia" (Buber, 1958, p. 8). In this vein, we might suggest that if salvation will come in "no-time" (never), then utopia will appear in "no-place".

One of the central questions is whether utopian thought should consider the condition of each individual separately, or whether utopianism is driven by the desire to establish a just society—even at the cost of harming individuals. According to another approach, utopianism should be seen as perpetual, internal reformation—almost detached from the outside world, although there is no reason that internal and external reformation cannot go hand in hand. The combination of internal and external reformation may seem less violent, but it in fact combines technology for the transformation of the world and technology for the transformation of "human nature". This kind of mutual reformation involves the dismantling and reconstruction of individuals and their surroundings. Utopia thus appears to harbor an internal contradiction, from which dystopia derives—like the snake in the Garden of Eden.

The discourse of progress that links scientific development and technological improvement to the Enlightenment project that aspires to social justice, democracy, and freedom is thus fundamentally utopian. When it comes to the nature of the process, however, the question remains whether it is possible to slowly move toward a utopian state, knowing full well that we will never get there and that we do not even fully understand what it means, or whether utopia must be pursued in the here and now, in the belief that it is only by means of a great leap, thoroughly eradicating evil in all its forms, that we may reach the promised land. When we think of utopia in terms of a great leap, it is necessarily eschatological, because the aspiration to bring it about in the here and now renders the matter of cost irrelevant: when we seek to turn no-place into here and no-time into now, the result is dystopia.

If utopia is the good place that does not exist, dystopia is the bad place (δυσ/*dus*, bad) that very much does exist. Claeys (2017) has shown that dystopia represents the failure to establish a utopia—the classic example of which is the socialist fantasy that turned into Stalinism. Whether dystopia is an accidental or necessary (or perhaps contingent—neither necessarily right nor necessarily wrong) outcome of utopianism is a fundamental question. In order to answer this question, we must remember Lenin's famous remark that you can't make an omelet without breaking eggs, since it removes the illusion that Stalinism was an accidental result of socialism. When we think of the figure of Leon Trotsky who, in many ways, saw himself as the Messiah of his time and thought in terms of "permanent revolution", it does indeed seem to be inherent to the socialist vision. Thus, Stalinism too may be inherent to socialism: exile, assassination, and mass murder are not divorced from utopian-eschatological thinking.

It is interesting in this context to note the words of Austrian-British philosopher of science Karl Popper:

> But of all political ideals, that of making the people happy is perhaps the most dangerous one. It leads invariably to the attempt to impose our scale

> of "higher" values upon others, in order to make them realize what seems to us of greatest importance for their happiness; in order, as it were, to save their souls. That is to say, it leads to Utopianism and romanticism. We all feel certain that everybody would be happy in the beautiful community of our dreams; and no doubt, there would be heaven on earth if we could all love one another. But … the attempt to make heaven on earth invariably produces hell. (Popper, 1963, p. 237)

While apocalyptic thought detaches the individual from the course of history, utopian thought seeks to restore individual power and ability to the course of history: "For eschatology the decisive act happens from above … for Utopia everything is subordinated to conscious human will" (Buber, 1958, p. 8). Upon closer examination, however, we discover a deep and inherent link between the apocalyptic state of mind and utopist fantasies that imagine the perfect society, detached from the individual. In both cases, the cost, in terms of human life and in terms of the hell on earth created on the path to a perfect society is practically unlimited. Thus, both apocalypse and social utopia appear to dismiss the individual.

In "*Tikkun Olam*: Between Utopian Idea and Socio-Historical Process" (Sagi, 2015), Israeli philosopher Avi Sagi suggests a distinction between two types of *tikkun olam* (lit. repairing the world; action to improve the world and make it a better place): *tikkun olam* as an idea and *tikkun olam* as a process. When understood as an idea, the longing for a better world rejects the current world; what should be supplants what is. When understood as a process, better possibilities are sought that do not reject the present but are, rather, embodied in it as realizable potential, while accepting human complexity. The totalizing ideal is essentially abstract and absolutist. It fails to see reality in shades of gray, acting in terms of black and white, all or nothing. It cannot tolerate almost, on the way, not yet, soon. It refuses to view reality through the complexity of the present or existing constraints, and is willing to sacrifice the present for the sake of a future program that ignores the individual. It rejects the existence of different notions of happiness. The processual approach, on the other hand, allows us to identify injustices in the present and rectify them even without the idea of a perfect world. Rather than a preconceived product to be aspired to, it offers a concrete future that can be shaped: "An essential and fundamental distinction must be made between utopias that come to be in the abstract and in the concrete. [Abstract utopias are] … not mediated with the existing social tendency and possibility … [and do not] … appeal to contemporary history" (Bloch, 1970a, pp. 89–90). The processual approach prefers hope and experience to ecstatic joy and absolute justice, which, according to French-Jewish philosopher Emmanuel Lévinas, always comes at the expense of the other, damaging the intimate bond between the other's wounding, demanding gaze and myself. It sanctifies non-indifference to the human Other, while the ideal approach—in the name of absolute justice—displays indifference to the plight of the individual. The process believes in this world and hopes to change it, viewing the future as flexible and subject to change. It prefers the touch of a nearby hand to the horizon-ward gaze, which tends to forget the world in favor of the romantic but violent sunset. Utopianism, as Milan Kundera explains in *The Unbearable Lightness of Being*, is kitsch—the logical conclusion of which is the concentration camp. The future is not preordained, and in order to change it, one must act in the present, cognizant of its constraints, close to the world and to people, and fully aware of the fact that we do not know what is the good today, let alone what the good will be for future humans.

In *The Imperative of Responsibility*, Jonas argues that technologies have, in the present era, made utopian fantasies possible within the foreseeable future, thereby making dystopia a very real possibility as well. Indeed, technology enables the realization of eschatological vision in the form of a secular utopia—dystopia in a Godless world. The more advanced the technologies, the more realizable even extreme possibilities become and the more palpable the danger they present. As we have seen in the case of the atom, technology is tempting and sweet. The possibilities that technoscience produces make incaution—due, in part, to the inherent difficulty in evaluating their true danger—a recipe for disaster. One of the gravest mistakes, according to Jonas, is the belief that there will be a technological breakthrough that will provide new solutions. For Jonas, this is too great and too dangerous a gamble. Therefore, in the absence of humility, technological utopia is in fact dystopia. According to Jonas, as one who sought to establish ethics as a basis for action in the current era, although the people of the future, whoever they will be, have no lobby to protect them from our fantasies in the present, we do not have the privilege to ignore the dangers inherent to the utopian-technological vision. We must, therefore, choose paths that may not lead to the ultimate good, but will certainly not lead to the ultimate evil.

The twentieth-century German-Jewish Marxist philosopher Ernst Bloch, considered a leading exponent of utopian thought and known primarily for his book *The Principle of Hope* (written during the Second World War), thought in terms of *Homo utopicus*—the expectation of a good future being inherent to human beings who ask the questions "where to" and "what for", a fundamental and primal element of our makeup: "To be human really means to have utopia" (Bloch, 1970b, p. 239). Bloch thus replaces Heidegger's Being-toward-death with a soberly optimistic not-yet:[1] "optimism with a black armband" (Bloch, 1970b, p. 241). The world, according to Bloch, is a laboratory that allows an infinity of possibilities for the construction of a better future, and, in this sense, humanity and the world are works in progress: "human content that is not yet clearly defined, not yet manifest: a concrete-utopian human content" (Bloch, 1970a, p. 144). Yet, according to Bloch, utopia lies not in prophecy or fantasy, but in revealing the real possibilities in the present to create a better life—not perfect, but better. It is an endless search for constant improvement in the present. Bloch rejects the ideal, Platonic utopia that is detached from the here and now, and sanctifies the unholy utopia that is concerned with human freedom, that understands the tension between justice and freedom, and that thinks of time in human terms: "The unique point of convergence of past, present and future cultures is a human content that is nowhere as yet adequately manifest, but can certainly be appropriately anticipated" (p. 144). Although Bloch indeed seems to propose a utopian possibility, Jonas (1984) rejects the notion of a utopia that is not eschatological and, at least in this era, of an eschatological utopia that is not technological. One might even say that no utopian thought can ever be sufficiently modest and adequately cautious—that is, a utopia that both nature and humanity can withstand.

Popper identified ideal-totalitarian utopian thought in Plato's *Republic*: the ideal state, according to Plato derives directly from the belief in "one absolute and unchanging ideal" (Popper, 1963, p. 161). Popper describes Plato's approach as "utopian social engineering", whereby we must first define "our ultimate political aim, or the Ideal State", and "only then can we begin to consider the best ways and means for its realization" (p. 157). According to Popper, "The Utopian attempt to realize an ideal state,

[1] Interestingly, Jonas rejects this utopian approach on a Heideggerian basis. Jonas's position is rather extreme: all utopias are essentially eschatological. The ethical challenge becomes particularly difficult at this point: we must safeguard the possibility of human existence in the future in a way that does not come at the expense of the present. What is more, we must do so not from the perspective of what the present might consider a good and worthy future, but out of a sense of openness to every possibility.

using a blueprint of society as a whole, is one which demands a strong centralized rule of a few, and which therefore is likely to lead to a dictatorship" (ibid., p. 159). One of the problems in an ideal state is that it will never achieve its aims: "The situation must become even worse for the Utopian engineer" (ibid., p. 160). Utopia is thus no less than apocalyptic. Precisely in this context, it is worth examining the Enlightenment vision.

The Enlightenment Concept of Utopia: The Desire to Dominate the World and Humanity

Following Socrates, the thinkers of the Enlightenment viewed ignorance as the source of evil: a person cannot know the good and still choose the bad. Accordingly, happiness, justice, and goodness are served by the war against ignorance and efforts to disseminate knowledge. The Enlightenment links technological and scientific progress to social justice. In the twentieth century, this idea was turned on its head. The German philosopher and sociologist Max Horkheimer argued that the dialectic of Enlightenment "consists largely in the change from light to darkness" (Horkheimer, 1978, p. 180).

The motto of the Enlightenment, in the words of Kant, is: "Have courage to use your own understanding". The Enlightenment philosophers' point of departure was that independent thought and reason (the Enlightenment is also known as the Age of Reason) are necessary and sufficient conditions for a better world. For Horkheimer, however, thought and reason do not necessarily promote the good; "Thought serves any particular endeavor, good or bad" (Horkheimer, 2004, p. 7). Under the rule of formalized and techno-scientific reason, we replaced thought with calculation and planning, and truth with propositional calculus. The result is the instrumentalization of reality, expressed not only in the lack of ability to distinguish between good and evil, but also in the elimination of these concepts and their replacement with more "enlightened" terms, such as efficiency, progress, power, and domination.

The modern intellectual, who renounced God but not divinity and replaced religion with science, never broke free of myth. On the contrary, rationality itself is the greatest myth of all. The modern intellectual is thus left with nothing, since reason and rationality do not seek the good, but domination: "Technical rationality today is the rationality of domination"

(Horkheimer & Adorno, 2002, p. 95). This is, perhaps, Horkheimer and Adorno's central argument.

The thread that links Socrates to Kant and the thought of his time is rationality. We believe that rationality will help us discover great truths and ultimately bring us to the gates of paradise. Rationality, however, "pertains to domination, not reason" (Horkheimer, 2004, p. 102). According to the French philosopher Michel Foucault, the Enlightenment motto "Knowledge is power" should be understood as follows: "The formation of knowledge and the increase in power regularly reinforce one another in a circular process" (Foucault, 1995, p. 224). Knowledge allows us to dominate—first our environment and then human beings themselves: "What human beings seek to learn from nature is how to use it to dominate wholly both it and human beings" (Horkheimer & Adorno, 2002, p. 2). According to this approach, at the very heart of the Enlightenment project lies the (negative) will to power and the desire to dominate the world and humanity (what Heidegger calls the will to will).

The Enlightenment is a never-ending project, in which human beings create themselves. History teaches us that it is not easy, and is often dangerous to create a new human, or, in the words of Horkheimer: "Advance in technical facilities for enlightenment is accompanied by a process of dehumanization. Thus progress threatens to nullify the very goal it is supposed to realize—the idea of man" (Horkheimer, 2004, p. v). The temptation to become a superhuman, who is beyond good and evil but simply acts based on the will to power is always there, in the background, inherent to the Enlightenment project. Indeed, after the twentieth century, the idea of "a new man" and "a new society" is not a promise but a threat:

> In fact we know from experience that the claim to escape from the system of contemporary reality so as to produce the overall programs of another society, of another way of thinking, another culture, another vision of the world, has led only to the return of the most dangerous traditions. (Foucault, 1984, p. 46)

As I have shown in my book, *The Uncanny and Technology: In an Effort to Imitate the Act of Creation* (Ataria, 2024), the current project of artificial intelligence works within the same framework—the desire to create a superhuman independent of perishable matter and free of the sins of the flesh—and is, at the same time, a project to destroy humans as we know them. According to Heidegger, the sole goal of humanity of the current era, that is, humanity that has already fallen into technological being, is to create an android: "Sometimes it seems as if modern humanity is rushing headlong towards this goal of *producing itself technologically*" (Heidegger, 2006, p. 197). It is important to understand that every project of renewed creation is, at the same time, a project of elimination of the previous species: "The new and improved forms of life tend to supplant the old and unimproved forms" (Darwin, 1998, p. 228).

Frederick the Great (1712–1786), King of Prussia, who promoted the ideas of the Enlightenment wrote that, "[The Enlightenment] was Prometheus who brought the heavenly light to earth in order to enlighten the blind, the people" (quoted in Ohana, 2019, p. 6). Some have suggested, however, that Prometheus sought to replace one divinity with another—that of humanity. Friedrich Nietzsche, the German philosopher (a limiting concept for Nietzche) active in the late nineteenth century who largely shaped discourse in the twentieth century, believed that the Promethean project is the will to power (in its negative sense) and the desire to dominate—which can lead to catastrophe. Based on this idea, Horkheimer wrote, "Enlightenment, understood in the widest sense as the advance of thought, has always aimed at liberating human beings from fear and installing them as masters. Yet the wholly enlightened earth is radiant with triumphant calamity" (Horkheimer & Adorno, 2002, p. 1).

In *Dialectic of Enlightenment*, which Horkheimer and Adorno began to write during the Second World War and published 2 years after its conclusion, the authors argue that there is a deep connection between the Enlightenment and the apocalypse of the Second World War: "The curse of irresistible progress is irresistible regression" (p. 28). They assert that the barbaric nationalism that emerged in Germany was a product of the Enlightenment: "The cause of enlightenment's relapse into mythology is to be sought not so much in the nationalist, pagan, or other modern mythologies concocted specifically to cause such a relapse as in the fear of truth which petrifies enlightenment itself" (p. xvi). It is important to understand that this was not a chance development: "[The] tendency toward self-destruction has been inherent in rationality from the first" (p. xix).

Horkheimer and Adorno are not the only ones to link Enlightenment and the Holocaust. In his book *Modernity and the Holocaust* (2008), the Polish-Jewish sociologist Zygmunt Bauman shows how there is, indeed, a deep—one is almost tempted to say necessary—connection between Enlightenment and modernity and the concentration camps. The

Holocaust was not the result of anti-Enlightenment. On the contrary, there is a direct link between the scientific revolution, rational thought, the mechanization of nature, the obsession with cataloging, comparing, and ranking, the engineering/medical division of the world into healthy and unhealthy elements that compromise the survival of the species, and the concentration camps. The apocalypse of the Second World War is a product of Enlightenment vision—of utopian thought.

The attempt to create a new reality is an apocalypse, because the act of creating a new human or a new world entails destroying the old world and its inhabitants. The theory of evolution lent scientific support to this idea:

> Extinction and natural selection will, as we have seen, go hand in hand … The extinction of species and of whole groups of species … almost inevitably follows on the principle of natural selection … Thus, from the war of nature, from famine and death, the most exalted object which we are capable of conceiving, namely, the production of the higher animals, directly follows. (Darwin, 1998, pp. 141, 384, 396)

Thus, Enlightenment, more than a breakthrough in Western thought is, in the words of Joseph Dan, the belief that "the end-time is not absolutely fixed, but depends, at least to some extent, on human beings, whose actions can hasten or delay it" (Dan, 2000, p. 56). This approach lies at the heart of the utopian aspiration to create a superior human. If the obstacle before creating a new human is human nature, then change must be effected on the genetic level. It is therefore worth examining the connection between techno-science and utopia as a real possibility.

New Atlantis: "Enlarging the Bounds of Human Empire, to the Effecting of All Things Possible"

Dan argues that in order to understand the utopian idea we must understand how religion and science converge—not as two opposing vectors, but as a single, metaphysical framework (Dan, 2000). As we have seen, the "discovery" of America played a central role in More's ability to imagine utopia. To the changing world map, we should also add changing modes of thinking and understanding, stemming from scientific thought. As philosopher of science Steven Shapin has stressed (Shapin, 1998), the terms "breakthrough" and "revolution" are inappropriate, and yet, one of the pivotal moments in what is generally called the "scientific revolution" is the 1543 publication of astronomer and mathematician Nicolaus Copernicus's *De revolutionibus orbium coelestium*. In this book—published, as legend has it, on the day of its author's death—Copernicus argues that the Earth is not the center of the universe.

The "discovery" of America as a real possibility to start anew and turn a no-place into a real place, and the scientific development that turned no-time into possible time, made utopia possible, at least in theory. It thus comes as no surprise that "many chapters in utopian literature express the desire to apply the new achievements of science … to the rectification of humans, as individuals and as a social group" (Dan, 2000, p. 188).

English philosopher Francis Bacon's novel *New Atlantis*, first published (in Latin) in 1624, tells the story of Salomon's House, "the noblest foundation, as we think, that ever was upon the earth, and the lantern of this kingdom. It is dedicated to the study of the works and creatures of God" (Bacon, 2020, p. 21). The goal of the project was not merely to learn the "causes, and secret motions of things", but also "the enlarging of the bounds of human empire, to the effecting of all things possible" (ibid., p. 33). Indeed, in the context of the will to dominate environment, which is an integral, perhaps inherent part of the Enlightenment vision, Salomon's House promotes the domination of plants and animals and seeks to extend human life. Bacon considered the promotion of science a religious duty, both as a way to increase knowledge of nature and thereby of the Creator, and as a way of helping others, by means of science's domination of nature. The role of Salomon's House is thus to reveal the "true nature" of nature and to bring people closer to Scripture, which Bacon viewed as the source of "natural Christianity". The role of scientists in this mission is thus to reveal the origins of religion (by revealing nature) and to convey this truth directly to the people, without intermediaries.

Bacon's 1620 book *Novum Organum* (New Tool) defined the empirical scientific project. The "new tool" was not a telescope or other technology, but a method: relying on empirical evidence and progressing by induction (from the particular to the general), in an effort to reveal basic axioms or natural laws.

Bacon declared that faith and science are ways of improving man and reality in our own lifetimes, in an attempt to establish better relations with the Creator. Bacon was a utopian because he believed that humans can change their fate, and, as Burdett suggests, was motivated by the eschatological tension between "even now" and "not yet" (Burdett, 2014). The scientific project seeks to change humanity in order to return to the bosom of God—to paradise lost. It is precisely in this aspiration that Jonas identifies the danger of scientific utopia (Jonas, 1984).

For Popper (1963), the messianic apocalypse is at one extreme, while rationalism and science are at the other extreme. Dan, on the other hand, is of the opinion that the relationship is better framed as the tension between belief in salvation and belief in improvement and rectification. He argues that ever since the atomic bomb, we have tended to ascribe "negative apocalyptic traits to the development of science [which] includes the possibility of creating a utopia, but, to the same extent, includes the dangers of a holocaust" (Dan, 2000, p. 266). Jonas goes further, and claims that apocalypse is inherent to the "Baconian program", that is, the technoscientific project (Jonas, 1984).

Dan draws our attention to the fact that at the root of rational thought and the desire to impose the rule of reason lie the seeds of totalitarian thought. In any event, "Not one of the components that create the totality of the apocalypse of reason negates, in and of itself, the use of violence" (Dan, 2000, p. 281). According to this approach, the scientific revolution and the Enlightenment vision on the one hand, and the utopian-messianic project on the other are merely two sides of the same coin—a single vector, and the desire to improve humankind is a prime example of this.

Eugenics: A Wonderful Homogeneous World

Utopian discourse takes us to the concept of "eugenics", coined by Darwin's cousin Francis Galton in 1883. Eugenics developed on the basis of Darwin's observations regarding the domestication of animals—from chickens to pigeons, cows, and dogs. Domestication involves human intervention in the reproductive process of animals, to cater to human needs.

The word eugenics derives from the Greek *eu* (εὖ), good, and *genos* (γένος), race, stock, kin, species. Combined, the two words create the expression "well born" or "of noble heritage". The concept itself reveals one of the main problems with this approach: the idea that there is some basic essence that is defined, a priori, as good. Evolutionarily speaking, this concept is fundamentally mistaken.

Not surprisingly, the authors of the theory, as members and representatives of a particular society, reduced the concept of "good" to that which they themselves considered good and desirable at the time. The problem that emerges (once again) from discussions of eugenics lies in the attempt to exploit belief in science in order to lend validity to a concept that in fact reflects the desiderata of those who happen to be in power—a further example of the fact that the concept "natural" is always political (as well).

The theory was also scientifically flawed, because, from an evolutionary perspective, the concept "good" has no clear or constant significance—certainly not of a predetermined nature. There is no way of knowing, a priori, what traits are better suited to the survival of the species. Evolution speaks of suitability not values. Thus, in evolutionary terms, the idea "good" merely refers to the ability of a particular species to survive (i.e., reproduce) under certain conditions. It is, of course, impossible to know in advance what traits a particular species may require to survive in conditions that may arise over the course of a lifetime. Dramatic changes in habitat are not necessarily the exception.

Eugenics, as a scientific approach, developed in the political, social, ideological milieu of nineteenth-century Victorian British society. It can be said, in general, that Victorian society accepted as practically self-evident, the division into two main classes: the poor and ignorant masses of the lower class, and the educated, wealthy upper class. Victorians also believed that science would ultimately resolve the central problems of the human race. These two elements are crucial to understanding the popularity of eugenics at the time.

The rigid division into bad-poor/rich-good was treated (almost) like a scientific fact in Victorian society. Viewed in this light, it is our origins that determine our development. In practice, this belief made it impossible to cross over from the lower to the upper class. It thus becomes clear why eugenic discourse was rooted both in the possibility of improving the ignorant and morally corrupt masses through the "manipulation of the laws of heredity", and in the fear that the welfare state would allow "deficient individuals who would otherwise have gone extinct, to remain alive and even to reproduce … Medical innovation interferes with the process of survival of the most worthy" (Graur & Graur, 1992, pp. 23–28).

Even today, similar ideas arise from time to time, in the context of "selfish gene" theory. Dawkins, one of the most important figures in the field of popular science in general and that of evolution in particular, claims that "Uncontrolled birth rates are bound to lead to horribly increased death rates … Humans (some of them) have the conscious foresight to see ahead to the disastrous consequences of over-population." Since genes are selfish, they "most certainly cannot be expected to see into the future, nor to have the welfare of the whole species at heart" (Dawkins, 2006, p. 144). For Dawkins, "Individuals who have too many children are penalized, not because the whole population goes extinct, but simply because fewer of their children survive." He continues, "Genes for having too many children are just not passed on to the next generation in large numbers, because few of the children bearing these genes reach adulthood" (Dawkins, 2006, pp. 151–152). The problem, according to Dawkins, is that in the context of a comprehensive welfare system, "the state, which means the rest of the population, simply steps in and keeps the surplus children alive and healthy". Dawkins believes that "the welfare state is a very unnatural thing … There is no welfare state in nature." The obvious conclusion, according to Dawkins, is that the welfare state should be limited. This conclusion and the way in which he arrived at it, ostensibly on the basis of scientific research, is not new.

Herbert Spencer, one of the most influential figures in nineteenth-century Britain, believed that evolution and progress are synonymous, meaning that evolution has a purpose. Spencer argued, ostensibly on the basis of Darwin's theory of evolution, that any government intervention that seeks to support the individual is "a curse disguised as a blessing", and therefore, as philosopher of science Oren Harman put it, "Unfettered competition alone would lead to the advance of civilization in the long run" (Harman, 2011, p. 26). Such views lie at the heart of extreme capitalism, and it is easy to see how, with only slight manipulation, these ideas allow those who are perceived as weak to be harmed "naturally".

The problem with this idea is not merely a moral one: "weak" is often a cultural, rather than biological concept, and we do not know (nor can we know) who will be better "suited" than others to survival in changing circumstances in the future. Here also lies the great absurdity of attempts to play with the genes of a child not yet born. And once again, the political use of the concept "natural" or "biological" may take us to dangerous places.

Eugenics is profoundly rooted in racist discourse, and provides tools for the application of racist ideas. Eugenics offers two main ways in which to shape society: purification and preference. The purification approach prohibits individuals of a certain type from reproducing, in order to "eliminate the targeted trait from the population" (Graur & Graur, 1992, p. 32)—either by eliminating the bearers of the undesirable trait themselves or merely preventing them from reproducing. The preferential approach acts to increase the incidence of a desirable gene in the population, by encouraging the reproduction of chosen individuals. And indeed, in the twentieth century, countries like Canada, Singapore, and the United

States enacted laws supporting either the purification or preferential approach. In Sweden, for example, researchers allowed the "voluntary" sterilization of the mentally ill, although the country brought significant pressure to bear on individuals, resulting in some 60,000 sterilizations in the years 1934–1974.

Eugenics is in deep inner conflict. On the one hand, it seeks to improve weaker groups; on the other, elite groups require legions of ignorant individuals to perform menial tasks for them and keep the society running. This explains why, at a certain stage, discussions of improvement gave way to discussions of suitability—the creation of human beings of various kinds, suited to various tasks.

In both the Soviet Union and Nazi Germany, utopian thought and eugenics were linked to one another. The Nazi regime employed sterilization and elimination (purification)—from the well-known T4 program (euthanasia)[2] to the elimination of babies born with a rash. This process was enshrined in the Law for the Protection of the Hereditary Health of the German People. At the same time, the Nazi regime encouraged the birth of "suitable people" in the belief that only in this way could they create a strong, able-bodied, and healthy nation. In the Soviet Union, the genetic approach was abandoned in favor of an environmental-influence-based approach, adopting the ideas of Jean-Baptiste Lamarck (1744–1829), considered the first natural scientist to develop the idea of evolution as a function of environmental factors. Lamarck argued that acquired characteristics could be passed onto the next generation. The socialist view, in its communist iteration, was that changes in the population would come about through a process of re-education and would be passed onto the next generation. The emphasis in communist society was thus on environmental factors: there is no such thing as the gene of a thief; it is the environment that creates the thief.

Bill Hamilton, one of the foremost evolutionary scientists of the twentieth century, argued that humanity faced two choices: to make eugenics

[2] In the Hitlerian process of purification, prior to the murder of the Jews and in parallel to the Polish campaign, Germany launched the T4 euthanasia program (Ger. Aktion T4), during the course of which some 90,000 Germans suffering from mental disorders, congenital abnormalities, and so forth were murdered (the project lasted from October 1939 to August 1941). Use of the term "euthanasia" was neither coincidental nor simply cheap demagoguery. It reveals a deep truth about the nature of utopian thought. At the heart of collective utopia lies a wall and an impossible desire for purity. In this sense, collective utopia is necessarily a dystopian reality for other groups.

as a more-or-less personal choice, or to become, through institutional repression, "mere executive cells within superorganisms" (Hamilton, 1996, p. 193). The second possibility is more familiar to us, because we have seen it before. It is a combination of the Nazi and Marxist-communist approaches. Hamilton stresses that there is another possibility, which he considers less likely, but that seems to be what we are currently witnessing. In the context of liberal capitalism, eugenics by choice appears to be happening. It is by no means clear that "liberal eugenics", as philosopher and sociologist of industrial and technologically developed, capitalist democratic societies Jürgen Habermas calls it, does not lead to dystopia. According to Habermas, parents who decide to reject certain genetic traits are not only preventing their future realization, but are also giving expression to and replicating a revulsion at parts of themselves (Habermas, 2003). Liberal eugenics thus brings about the annihilation of the other within us—eliminating the Hell that is us. Dystopia in the case of liberal eugenics is the result of an inability to bear even ourselves. At the heart of liberal eugenics lies self-loathing and the desire for self-destruction, which is, as we have seen, apocalyptic desire.

Although liberal eugenics is based on science, its very approach—rational planning—is the exact opposite of the principles of evolution. While evolution is essentially the result of blind mutations and selection, attempts to shape human nature lie within the realm of intelligent design—God as a super-engineer returns in the form of the planning scientist. The influential historian Yuval Noah Harari put it as follows: "Biologists the world over are locked in battle with the intelligent-design movement ... [They] are right about the past, but the proponents of intelligent design might, ironically, be right about the future" (Harari, 2014, p. 448). Worse, such efforts at intelligent design are funded by for-profit companies, so that this planning is ultimately the product of a socioeconomic dynamic that, like nineteenth-century Victorian society, defines concepts of "good", "bad", "suitable", "weak", and so forth in very narrow terms. Philosopher of science Snait Gissis explains:

> Genetic engineers present themselves as engaging in therapy, repair, improvement, and equal opportunity. Their focus, however, is on individuals and therefore entails a fundamental disregard for conditions of severe economic disparity between and within societies, and thus fails to address the ramifications of these conditions of disparity on the application of the practice and its future results. The rhetoric helps disguise the fact that

> *the practice of genetic engineering purports to offer a biological solution for ills that are, at least in part, social ills*—the result of internal and global class inequality, evident in the distribution of material, cultural, and medical resources. (Gissis, 2020)

Intelligent design attempts to determine the present in a changing world and harms "natural" evolutionary processes, without knowing what impact a particular genetic modification might have on the organism as a whole, and without knowing what constitutes "suitability" in a future world we cannot even imagine.

The liberal eugenics project, like any project that tries to shape human nature, is only partially scientific. Such efforts at design should be considered within the framework of the broader goal, inherent to the Enlightenment project, of dominating human nature. From this perspective, knowledge is indeed power—the power to control and dominate.

A deep understanding of the concept of eugenics explains the inherent connection between utopia and dystopia. Dystopia is not a chance result of utopia. Utopia and dystopia are not even two sides of the same coin, but a single concept, the very same phenomenon. A society that rejects diversity and the other—not merely as an idea but in practice—is a society that dictates, directly or indirectly, who is worthy of life and under what conditions. Utopia is dystopia because there is no utopian project that does not include large-scale eliminations. Utopia is dystopia because the attempt to achieve perfection entails the murder of those who are deemed unsuited and, as we have seen, the concept of "suitability" changes in keeping with the needs of those in power.

A Brave New World

George Orwell's *1984* and Aldous Huxley's *Brave New World* are the two great dystopian novels of the twentieth century. In Orwell's dystopia, the means to achieving the ultimate goal is terror and violence—which is not the case in Huxley's novel. Following the book's publication, in 1949, Orwell sent a copy to his French teacher, Huxley, who wrote the following in response:

> Whether in actual fact the policy of the boot-on-the-face can go on indefinitely seems doubtful. My own belief is that the ruling oligarchy will find less arduous and wasteful ways of governing and of satisfying its lust for

> power, and these ways will resemble those which I described in *Brave New World* … Within the next generation I believe that the world's rulers will discover that infant conditioning and narco-hypnosis are more efficient, as instruments of government, than clubs and prisons, and that the lust for power can be just as completely satisfied by suggesting people into loving their servitude as by flogging and kicking them into obedience. In other words, I feel that the nightmare of *Nineteen Eighty-Four* is destined to modulate into the nightmare of a world having more resemblance to that which I imagined in *Brave New World*. The change will be brought about as a result of a felt need for increased efficiency (Huxley 1969, pp. 604–605)

In *Amusing Ourselves to Death*, Neil Postman (1986)—one of the foremost American media theorists of the twentieth century—wrote that Orwell's violent vision, which is a clear critique of the communist regime, is indeed frightening, but the vision we are in fact well on the way to realizing today is that of Huxley—a pleasure-based dystopia. You must consume and you must be happy, or you have no place in the world.

The book *Brave New World*, the title of which is taken from Shakespeare's *Tempest*, is set in London, a city in the World State, in the year A. F. 632. The A. F. ("after Ford") reckoning begins with the introduction of the Ford Model T. Paradoxically, there are no cars at all in the book. The story's constitutive trauma is the Nine Years' War, which began in A. F. 141 and in which nearly all of humanity was wiped out. After the war, the world controllers came to the conclusion that fundamental change was required if the human race was to survive: no longer the Kantian autonomous individual at the center, but "Community, Identity, Stability". Being part of a stable community meant negating one's individual identity (like Orwell in *1984*, Huxley uses a play on words, in which the word "identity" in fact signifies "sameness").

Mustapha Mond, one of the most powerful controllers in the World State and a former scientist who, in his youth, was nearly exiled from the communal space for daring to think for himself, explains that the best way to realize the motto "Community, Identity, Stability" is not through the use of force, but by means of "the slower but infinitely surer methods of ectogenesis, neo-Pavlovian conditioning and hypnopaedia" (Huxley, 1968, p. 34).

The tools that Huxley presents are a combination of eugenics (in vitro fertilization) and an aggressive system of brainwashing (hypnopaedia), and conditioning (electric shock). The result is a class-based society in which everyone is happy with their lot: "That is the secret of happiness and virtue—liking what you've got to do. All conditioning aims at that: making people like their unescapable social destiny" (p. 10). The brainwashing persisted, "Till at last the child's mind *is* these suggestions, and the sum of the suggestions *is* the child's mind … The adult's mind too—all his life long" (p. 19). It is worth emphasizing the obvious: these are "Suggestions from the State".

At the Hatchery and Conditioning Centre, they produce 96 identical twins from a single egg—and constantly strive for more. Every embryo is genetically predestined for its specific caste, as an Alpha, Beta, Gamma, Delta, or Epsilon, and every child is raised in the appropriate environment: "An Epsilon embryo must have an Epsilon environment as well as an Epsilon heredity" (p. 9).

In the brave new world, people have neither mother nor father, brothers nor sisters, home nor family: "Every one belongs to every one else" (p. 26). There are therefore no intimate relationships, only frequent sex with a variety of partners. Despite the principle of community, no one cares when someone dies—one might even say that people do not die at all, but are merely replaced, that is to say, everything is reduced to a standing-reserve. Bernard Marx—a psychologist (specialist on hypnopaedia) and an Alpha Plus who is physically inadequate (eight centimeters short of standard Alpha height and slender) and who does not respond properly to conditioning—considers the whole thing little more than a meat market. Of course that is only when he is left out. The moment he himself is desired, he sees things quite differently.

People in the brave new world have no real age. For four decades, they look and act young, and then their systems suddenly shut down: "All the physiological stigmata of old age have been abolished … Now—such is progress—the old men work, the old men copulate, the old men have no time, no leisure from pleasure, not a moment to sit down and think … Youth almost unimpaired till sixty, and then, crack! the end" (pp. 37, 74). After they die, they are cremated—and not for ecological reasons. In the brave new world, there is no room for death. Death upsets stability.

With a few exceptions, the people Huxley describes are without personal identity and incapable of independent thought; "men as steady as the wheels upon their axles, sane men, obedient men, stable in contentment" (p. 28). Mond explains that excessive emotions threaten stability and obedience and therefore, "No pains have been spared to make your lives emotionally easy—to preserve you, so far as that is possible, from having emotions at all" (p. 29). Of course, what is liable to provoke unpleasant emotions is reality itself, which must therefore be repressed and denied. It is interesting, in this context, to note that Hannah Arendt, in her monumental *Origins of Totalitarianism*, contends that the totalitarian regime, inherently refuses to accept reality: "Totalitarian propaganda thrives on this escape from reality into fiction, from coincidence into consistency … The masses are obsessed by a desire to escape from reality" (Arendt, 1979, p. 352). Thus, even without the "boot-on-the-face", the new world is fundamentally totalitarian.

The brave new world, where happiness is an imperative, provides all of the necessary tools to escape from reality. In A. F. 184, *soma*, the perfect drug, was invented, described as "Euphoric, narcotic, pleasantly hallucinant … All the advantages of Christianity and alcohol; none of their defects … Take a holiday from reality whenever you like, and come back without so much as a headache or a mythology" (Huxley, 1968, p. 36). It almost goes without saying that the question that arises over and over again when reading the book is, *if everything is so wonderful, why is there such a demand for soma*?

Huxley describes a world in which the gravest danger is the possibility of thinking independent thoughts; a world in which slogans are constantly recited, in which there are rigid class divisions; a world in which happiness is an obligation, blindness and denial of reality a supreme value, and truth and science a threat; a world that rejects art: "That's the price we have to pay for stability. You've got to choose between happiness and what people used to call high art. We've sacrificed the high art" (p. 150). These are the elements that make the world Huxley describes a dystopia.

At this point I would like to ask, whether the dystopia described in *Brave New World* is not one-dimensional. While the world created by Orwell in *1984* is clearly dystopian, that is not the experience in Huxley's case. A sensitive reading of the book reveals another side as well. There are moments in which it is hard not to think that Huxley is actually describing a utopia. The ideal of returning to nature presented in the book is not a real option in A. F. 632. The Savage is shunned and isolated at the Reservation simply because he comes from a different background, because the color of his skin is different, and because his mother, a Beta Minus who was injured and left there, lives according to a different set of values. What is clear is that there is no acceptance of the other at the Reservation. The Savage, whom Marx brings to London, looks on civilization with disgust, while viewing Shakespeare as a kind of idyll. But was

England of the late fifteenth and early sixteenth-century, the period in which Shakespeare lived, close to a utopia? In 1600, the East India Company, symbol of European colonialism and racism, was founded.

After the Savage, so-named to evoke Rousseau's "noble savage", rejects civilization and progress, he decides to seclude himself. At this stage, we see him in his ultra-religious version. He whips himself bloody, like a Shiite practicing *tatbir* or an Indian Fakir. Is this return to guilt-ridden Christian monotheism a method? A utopia? Huxley is by no means naïve. He understands full well that longing for myths like that of the noble savage poses no less of a danger than falling into a state of total technological being, in the Heideggerian nihilistic sense (enframing).

When the Savage tries to free a khaki-clad crowd of 162 virtually identical Deltas by denying them their daily *soma* ration, which he calls "Poison to soul as well as body" (p. 143), he quickly abandons his high-minded words and passes to violence: "I'll teach you; I'll make you be free whether you want to or not" (p. 145). Is this the utopia of freedom that we so desire? And in what sense is the Savage himself free? He seems to be no less a prisoner of a radical belief system than the people of London.

In light of the above, I would ask once again, is Huxley's *Brave New World* a dystopia? When we read the book's sex scenes, the impression we get is that Huxley did not see orgies and sexual promiscuity only as a problem—at least not in all cases. It would be interesting to explore the extent to which men and women today aspire to this kind of sex and, some might say, actively try to achieve it. Other questions worth examining: how many of us would not wish to enjoy youth for 60 years and avoid the infirmity of old age? How many of us would not wish to love what we do and be happy in our jobs, even if it is the result of conditioning?

Some would argue that it is not these things that make *Brave New World* a dystopia, but the elimination of freedom. But is freedom preferable to happiness—even if it is false? *Soma* indeed makes one blind to the world, but don't people take drugs today too, and would we not wish to have the perfect drug to make us forget reality? Is it not true, as Postman argues, that current technological diversions are merely a version of *soma*? Ultimately, we must ask ourselves, would we not be willing to renounce freedom for the sake of the pleasure machine? I would argue that although many of us would answer in the negative, our behavior in this era has said yes, time after time.

Another sensitive point is that of the caste system in the *Brave New World*. I don't think anyone would be shocked if I were to say that even today there are immense class differences—between and within societies. Indeed, under the guise of climate discourse, contemporary elites find ways of justifying and perpetuating this inequality. The disgraceful argument goes as follows: if the lower classes begin to consume as we do, it will be the end of us all.

In any event, Mond presents some rather convincing arguments against egalitarian societies: "We believe in happiness and stability. A society of Alphas couldn't fail to be unstable and miserable" (p. 151). It is worth stressing that Alphas are no happier than the members of the other castes. On the contrary, says Mond, the lower classes are in fact the happiest. In support of this claim, Mond cites an experiment conducted in Cyprus in A. F. 473. After having cleared the island of its previous inhabitants, 22,000 Alphas were settled there. Within 6 years, a civil war had broken out and most of the colonists were killed.

It is somewhat surprising that Mond knows anything at all about past events, since one of the best-known lines in the book is "History is bunk"—based on words that Henry Ford actually said in 1916 ("History is more or less bunk"). But who is to say that historicism—the theory that history has a power and mission of its own, or, in the words of Joseph Dan, "an internal necessity stemming from the preordained nature of history … which individuals are powerless to influence … for good or for ill" (Dan, 2009, pp. 54–61)—is any better than Ford's approach?

Ostensibly, Ford's approach frees us from the bonds of history. Let us examine this thesis. Mond knows history and learns from the mistakes of history, but is not bound by it. The view articulated by Ford (a racist and a publisher of anti-Semitic propaganda) is the most dangerous of all. Just because we relate to history as if it were bunk doesn't mean that it ceases to control our lives—as Heidegger realizes, history is not in the past, but shapes the present. For the purposes of the present discussion, we may consider Ford's ideas irrelevant or even bunk, but we cannot ignore the fact that all American industry was created in the image of Henry Ford, whom *Fortune* understandably named Businessman of the Twentieth Century. Ford's vision of a car in every garage shapes Western perceptions of the automobile to this day. We need not know about Ford, because history is of course bunk, but that does not mean that history does not shape the present. In fact, under the guise of liberation from history, Ford presented a model of absolute slavery to history.

Mond's approach is more realistic. He learns history and understands how important it is for the reality of our lives in the present, since we are always, a priori, thrown into a world shaped by its history. Mond does not consider the new world all that wonderful, but is aware of the alternatives and finds it preferable to the world war and economic crisis precipitated by the previous system. It is a conscious choice between two bad options.

I would go a step further, however, and dare to suggest that Huxley's world is in fact wonderful—a true utopia, in the deepest sense of the word. Utopian discourse, as we have seen, was related to the "discovery" of America and the possibility of an island separate from the world, a place in which to begin anew. In *Brave New World*, islands play a central role—a clear allusion to More's novel, published exactly four centuries before Ford made his famous quip about history, under the title *On the Best State of a Republic and on the New Island of Utopia*.

Ostensibly, the islands (Iceland, Cyprus, Saint Helena, the Marquesas, Samoa, and the Falklands) are Stalinesque gulags, but in reality, they are nothing of the sort. As Mond puts it in a conversation with Marx, the Savage, and the third and most impressive side of the triangle—a rather talented Alpha Plus male by the name of Helmholtz Watson:

> One would think he was going to have his throat cut ... Whereas, if he had the smallest sense, he'd understand that his punishment is really a reward. He's being sent to an island. That's to say, he's being sent to a place where he'll meet the most interesting set of men and women to be found anywhere in the world. All the people who, for one reason or another, have got too self-consciously individual to fit into community-life. All the people who aren't satisfied with orthodoxy, who've got independent ideas of their own. Every one, in a word, who's any one. (Huxley, 1968, p. 154)

The islands demonstrate that it is not a dystopia. As Mond explains, "There are such a lot of islands in the world. I don't know what we should do without them. Put you all in the lethal chamber, I suppose" (pp. 155–156). The islands allow the existence of a kind of Eden of intellectual freedom within a society in which identity (in the sense of sameness) is sacred.

The book's message is thus far less simple than it appears. It is neither a utopia nor a dystopia—not because the reality it describes lacks utopian or dystopian elements, but for a far more basic reason: utopia and dystopia are one and the same. We cannot think of utopia but in relation to dystopia, because there can be no utopia without lethal chambers and crematoria. Therefore, any attempt to create heaven on earth, the Enlightenment's messianic aspiration, is necessarily dystopian. The goal of Enlightenment and science to create a new and worthy human, a new and just society, more than representing utopia attest to an apocalyptic mode of being—the desire to destroy ourselves.

Fragments of Salvation Made of Enriched Uranium

Utopia is an attempt to create fragments of salvation in the present. In order to realize utopia, we must first define "human nature", for only in that way can we know what type of change is required. But is a new reality, as in More's case (America) or in the case of Marxism, all that is required, or must we also improve our own nature—through eugenics and the elimination of inferior races and the physically and mentally disabled?

In order to hasten or delay the apocalypse, we must discover the bones of the first human. We must first understand what is flawed and only then can we define the problem in technical terms and devise some sort of final solution. To put it another way, considering human nature, are external forces sufficient, or do we need to change our internal systems in order to achieve progress (even voluntarily)?

In this part of the book, I have focused on utopian thought, because it seems to offer an explanation, albeit a partial one, for the bomb. Los Alamos is often described as a scientific utopia—an actual Salomon's House. Utopia is fundamentally total and cannot abide half-measures. In utopia, there is a need, an unstoppable urge, to reach perfection. The atomic bomb (and even more so the hydrogen bomb—a cleaner, far more destructive, and therefore more humane weapon) is the perfect product of a techno-scientific utopia, because it is essentially all-destroying—a total weapon. The atomic bomb is motivated not only by the desire to bring about doomsday, but is also a direct, or at least non-incidental product of a Bacon-style Salomon's House—in essence, what Truman himself wrote in his memoirs:

> The greatest marvel is not the size of the enterprise, its secrecy, nor its cost, but the achievement of scientific brains in putting together infinitely complex pieces of knowledge held by many men in different fields of science into a workable plan ... What has been done is the greatest achievement of organized science in history. (Truman, 1965, pp. 465–466)

The central thesis of this chapter is that this great marvel, the result of a utopian vision, is dystopic—and not by chance.

References

Arendt, H. (1979). *The origins of totalitarianism.* Harcourt Brace and Company.

Ataria, Y. (2024). The uncanny and technology. In *In an effort to imitate the act of creation (Heb.).* Pardes.

Bacon, F. (2020). In G. B. Wegemer (Ed.), *New Atlantis.* CTMS Publishers at the University of Dallas.

Bauman, Z. (2008). *Modernity and the holocaust.* Polity Press.

Bloch, E. (1970a). *A philosophy of the future* (J. Cumming, Trans.). Herder and Herder.

Bloch, E. (1970b). *Tübinger Einleitung in die Philosophie, Gesamtausgabe* (Vol. 13). Suhrkamp.

Buber, M. (1958). *Paths in Utopia* (R. F. Hull, Trans.). Beacon Press.

Burdett, M. S. (2014). *Eschatology and the technological future.* Routledge.

Claeys, G. (2017). *Dystopia: A natural history: A study of modern despotism, its antecedents, and its literary diffractions.* Oxford University Press.

Dan, J. (2000). *Apocalypse then and now (Heb.).* Sifre Hemed.

Dan, J. (2009). *Chaos theory and the study of history (Heb.).* Kinneret, Zmora-Bitan, Dvir.

Darwin, C. (1998). In G. Beer (Ed.), *The origin of species.* Oxford University Press.

Dawkins, R. (2006). *The selfish gene.* Oxford University Press.

Foucault, M. (1984). What is enlightenment? In M. Foucault, & P. Rabinow (Ed.), *The Foucault Reader* (C. Porter, Trans. pp. 32–50). Pantheon Books.

Foucault, M. (1995). *Discipline and punish: The birth of the prison* (A. Sheridan, Trans.). Vintage Books.

Gissis, S. (2020, May). The Racial assumption in genomic science (Heb.). *Hazman Hazeh.*

Graur, M., & Graur, D. (1992). *Eugenics: Improving the hereditary quality of man (Heb.).* Ministry of Defence.

Habermas, J. (2003). *The future of human nature* (W. Rehg, M. Pensky, & H. Beister, Trans.). Polity.

Hamilton, W. D. (1996). *Narrow roads of gene land* (Vol. 1). W. H. Freeman.

Harari, Y. N. (2014). *Sapiens: A brief history of humankind.* Vintage Books.

Harman, O. (2011). *The Price of altruism: George Price and the search for the origins of kindness.* W. W. Norton & Company.

Heidegger, M. (2006). In W. McNeill (Ed.), *Pathmarks.* Cambridge University Press.

Horkheimer, M. (1978). *Dawn and Decline: Notes 1926–1931 and 1950–1969* (M. Shaw, Trans.). The Seabury Press.

Horkheimer, M. (2004). *Eclipse of reason.* Continuum.

Horkheimer, M., & Adorno, T. W. (2002). In G. S. Noerr (Ed.), *Dialectic of Enlightenment* (E. Jephcott, Trans.). Stanford University Press.

Huxley, A. (1968). *Brave New World*. Bantam Books.

Huxley, A. (1969). In G. Smith (Ed.), *Letters of Aldous Huxley*. Chatto & Windus.

Jonas, H. (1984). The imperative of responsibility. In *Search of an Ethics for the Technological Age* (H. Jonas, & D. Herr, Trans.) Chicago University Press.

Ohana, D. (2019). *The intellectual origins of modernity*. Routledge.

Popper, K. R. (1963). *The open society and its enemies* (Vol. 2). Princeton University Press.

Postman, N. (1986). *Amusing ourselves to death: Public discourse in the age of show business*. Penguin Books.

Sagi, A. (2015). Tikkun Olam: Between utopian idea and socio-historical process. In A. Sagi, H. Tirosh-Samuelson, & A. W. Hughes (Eds.), *Existentialism, pluralism, and identity* (pp. 103–122). Brill.

Scholem, G. (1995). *The messianic idea in judaism and other essays on jewish spirituality* (M. A. Meyer, & H. Halkin, Trans.). Schocken Books.

Shapin, S. (1998). *The scientific revolution*. University of Chicago Press.

Truman, H. S. (1965). *Memoirs by Harry S. Truman* (Vol. 1). Signet Books.

CHAPTER 11

Conclusion

In the first part of the film *2001: A Space Odyssey*, we encounter apes who have no advantage over any other species in their environment. They are part of nature and in no way dominate it. We might even say they have no concept of nature, since they are completely immersed in it. After the first appearance of the monolith, we witness a breakthrough: the leader of the band of primates understands how a bone can be used as a tool. He begins by grasping the bone. In this sense, Kubrick begins his film from a Heideggerian perspective,[1] in which the act and the practice precede the theory:

> The interfusion of form and matter prevailing here is, moreover, controlled beforehand by the purposes served by jug, ax, shoes. Such usefulness is never assigned or added on afterward to a being of the type of a jug, ax, or pair of shoes. But neither is it something that floats somewhere above it as an end. Usefulness is the basic feature from which this being regards us.... [It] is always the product of a process of making. It is made as a piece of equipment for something. ... This name designates what is produced expressly for employment and use (Heidegger, 2008, p. 154)

[1] The use of the bone occurs after the primate stares at it for a few seconds. Only then does he begin to play with it. Arguably, in those seconds (which perhaps last an eternity), the primate makes a singular cognitive leap that renders action possible—in the sense of I think therefore I am. In my next book, which is on the subject of technology, I address this issue and show that tools came first and that they are, in fact, the evolutionary motor of the genus *Homo*.

Y. Ataria, *Genes, Technology, and Apocalypse*,
https://doi.org/10.1007/978-3-031-59197-6_11

In *2001*, we witness the transition from primitive tools (a bone) to HAL 9000—a supercomputer with perfect artificial intelligence that, during the course of a spaceship voyage to Jupiter, tries to kill all the astronauts on board (Fig. 11.1). In both cases, it is technology; but to what extent is it the same technology? French philosopher and anthropologist Bruno Latour wrote, "Many archaeologists proceed on the assumption that the *basic tool kit* (as I call it) and techniques are directly related by an evolution of tools into composite tools. But there is no direct route from flints to nuclear-power plants" (1994, pp. 61–62). Indeed, the gap between the bone and HAL 9000, perfect artificial intelligence—which, in light of technological developments, is anything but science fiction—is profound and essential. According to this approach, the passage from tool-use to modern technology should be described less in terms of evolution than in terms of the gap between the use of technology (tools of various kinds) and technological being: technology→**t**echnology. The film *2001* portrays this transition perfectly. Latour describes it as follows:

Fig. 11.1 Between Technology as a Tool and Technology as a Total Mode of Being
In one of the most famous scenes in the history of cinema—the throwing of the bone into the air, which, in a single motion, skips over millions of years of evolution—we pass from that bone (the first tool) to a space station (perhaps the last tool), indifferently moving in the void. It is thus a passage from the use of tools, that is the use of technology, to looking at the world from the perspective of technology, that is, the inability to see anything but technology in the world

> A huge bone lying at the water hole is suddenly seized by a rapidly evolving ape, transformed into a tomahawk, and used to break the skull of an enemy primate. (Tools and weapons, intelligence and war, commence all at once in this masculine myth.) The Promethean ape, thrilled by this invention and sudden change in the fortunes of war, launches the bone into the sky; the bone whirls around, then—again, suddenly—becomes a vast futuristic station, slowly turning on itself in the depth of space. From tools to high technology, millions of years are summarized in one beautiful cut. (ibid., pp. 42–43)

This scene raises the question of whether we controlled the bone to begin with and whether we control technology today. In other words, have we ever controlled technology or has that always been an illusion? Kubrick seems to think that it is, in fact, an illusion: from the bone to artificial intelligence, there is a certain continuity, an internal dynamic, and humans have never really controlled technology. This question has dramatic ramifications, because it pertains directly to the violent use the primate leader makes of the bone. Indeed, one of the most important and fascinating questions of our time is whether violence is inherent to technology or whether technology is neutral. In *2001*, Kubrick presents a clear position: when the leader of the band of apes, Moon-Watcher, learns to use a bone as a tool, it becomes a tool of murder. It is no coincidence that in the very next scene, the band is eating fresh meat and, as we saw in the first part of the book, cooperation has played a central role in human development.

The use of tools opens up new opportunities and redefines the individual and the band. Such an equation, almost by definition, nullifies the concept of human responsibility. Kubrick's approach—at least until the awakening of astronaut Dave Bowman when he is locked out of the ship—precludes our ability to control technology. For Kubrick, it is we who have been controlled by technology from the very beginning. Kubrick is thus an unwitting participant in the plan to absolve humanity of responsibility for Auschwitz and Hiroshima.

As we have seen, there is another way of interpreting this scene: Moon-Watcher is violent by nature; violence is in his blood. We know today that Kubrick was influenced, at least indirectly, by the work of Raymond Dart, discoverer of *Australopithecus africanus*, which he identified as an ancestor of modern humans. Dart characterized *Australopithecus* as a bloodthirsty ape whose entire development was a product of its need for blood. Like most people, Kubrick's familiarity with Dart's research was almost certainly through the popular works of Robert Ardrey, which place the violent, hunter male at the very center of the human evolution. Seen in this light, Moon-Watcher was violent and bloodthirsty long before he learned to use a bone as a tool. Technology did not change anything, but merely enabled him to carry out what he always had inside him. We are thus little more than male chimpanzees with advanced technology. Here too, humans are not to blame. It is in our genes, in our blood. It is stronger than we are—always was and always will be.

For Latour, humans and technology should be treated as equal partners: "The myth of the Neutral Tool under complete human control and the myth of the Autonomous Destiny that no human can master are symmetrical." He stresses, however, that "a third possibility is more commonly realized: the creation of a new goal that corresponds to neither agent's program of action. (You had wanted only to hurt but, with a gun now in hand, you want to kill)" (Latour, 1994, p. 32). Even according to this approach, however, humans are not to blame. It's part genes, part technology, but in any event, it is beyond our control.

2001 was released four years after another important Kubrick film: *Dr. Strangelove or: How I Learned to Stop Worrying and Love the Bomb.* During the course of the latter, we see again and again how Dr. Strangelove involuntarily performs the Nazi salute. At other times, he tries to strangle himself—once again, apparently involuntarily. The two gestures are mirror images of one another. The strangling hand is also the saluting hand, and the message is clear: the desire for self-destruction is inherent to human nature. The saving power of the apocalypse that Dr. Strangelove presents is embedded at the deepest possible level of Western culture.

When we examine the two films, *Dr. Strangelove* and *2001*, we understand that Kubrick is completely immersed in the apocalyptic mindset. Whether it is techno-science or the first human, the end is foretold: self-destruction is inherent to human existence. It seems, according to Kubrick, that humanity's end is inherent to its existence—beyond its control and, in many ways, beyond the realm of its responsibility. Kubrick did not invent anything, but merely exposed the archetypes that shape our culture and our way of being, as expressed in every aspect of human existence—from philosophy and science to history and popular culture.

If Kubrick represents the entertainment aspect of this approach, scholars like Dawkins, Diamond, and Harari represent the (pseudo-) scientific aspect—which are, in effect, two sides of the same coin. Indeed, despite their aura of respectability, Dawkins, Diamond, and Harari are not that different from Ardrey. Like Ardrey, they are thrown into the world (i.e., thinking, acting, and understanding) in the apocalyptic mode of being, which precludes the possibility of human rebellion. Indeed, between the selfish gene and domination by technology, it is unclear how much room, if any, is left for human will.

The success of these thinkers is due to the fact that they perfectly reflect humanity's current mode of being—that of the end-time. A good example of this is the title of Diamond's 2005 book *Collapse*, which leaves little to the imagination. Indeed, according to Diamond, although he might not admit it, there is a direct link between the collapse of the human race and its biological origins, as presented in his 1992 book *The Third Chimpanzee.* Diamond's work offers one of the most striking examples of the attempt to link the foundations of human evolution (reduction to the male chimp, with greater cognitive and cultural abilities) and a particular social structure that leads to self-destruction. As strange as it may sound, just as Heidegger identifies the tendency to self-destruction in technology, Diamond identifies the tendency to self-destruction at the heart of human culture. In this sense, the Nazi philosopher Heidegger, born (like Hitler) in 1889, and the Jewish-American biologist and geographer Diamond agree that there is a tendency to self-destruction at the very core of culture. As we have seen, Kubrick saw that very same tendency at the core of Western democratic regimes.

Whether it is the bones of a species we deem the missing link between chimps and humans, like Dart, Dawkins' selfish gene, or the technological dynamic advanced by Marx, Heidegger, and Elul, there is no room for humanity; the end is foretold. There is some comfort in that, however. If the end is preordained, we are free to destroy our world. Have no fear, we are well on the way: "It's not dark yet but it's getting there."

References

Heidegger, M. (2008). The origin of the work of art. In M. Heidegger & D. F. Krell (Eds.), *Basic writings: From being and time (1927) to the task of thinking (1964)* (pp. 139–212). Harper Perennial.

Latour, B. (1994). On technical mediation: Philosophy, sociology, genealogy. *Common Knowledge, 3*(2), 29–64.

Afterword

There is something wondrous about science, for me at least. I always want to know more about the universe, the Earth, or evolution, and I have a tremendous, unquenchable thirst to understand. Science looks like an immense and hallowed structure, in which there is order and perhaps even some certainty. Science, however, also excels at stealing our freedom of choice from us, but, to be honest, there is something pleasant and comfortable in giving up the ability to choose. It makes our lives much simpler.

It is somewhat disappointing to discover that scientists are human, and even more disappointing to discover that they are human first and foremost and, as such, are subject to the same basic archetypes as everyone else. They are neither better nor worse. This insight quickly leads to the understanding that scientists too are trapped in myths, and that we are unable to think outside of myths. The question then arises, what are the myths within which we operate?

Looking back on nearly 20 years of writing, I realize that this topic has fascinated me ever since my first book, *The Structural Trauma of Western Culture* (2017), which deals with the biblical story of the Binding of Isaac, which I define as the structural and constitutive trauma of Western culture. Indeed, the Binding of Isaac appears in countless works of art. Although this myth is very prominent in the development of the atomic bomb—reinforced by the name "Little Boy" given to the bomb dropped on Hiroshima—it is hard to see how it shapes day-to-day scientific work.

Y. Ataria, *Genes, Technology, and Apocalypse*,
https://doi.org/10.1007/978-3-031-59197-6

In recent years, I have taught a course on the link between trauma and culture.[1] Over the years, I have found myself talking in class about Auschwitz as the ground zero of Western culture—a foundational trauma that redefined the rules. Of course, the destruction of the Jewish people can also be described in terms of the *return of the repressed*: Christianity, the religion of the sons, returning to settle accounts with Judaism, the religion of the fathers.[2] In this sense, the Binding of Isaac remains relevant to our understanding of the dramatic events of the twentieth century, although, as I have already pointed out, the connection to contemporary science is a little more elusive.

* * *

When I began writing this book, I wanted to write a book for my son, explaining the fundamentals of current scientific approaches, and more precisely, how science accounts for the world as we know it, with an emphasis on evolutionary research. Time passed and the book took on a life of its own, practically writing itself, as if following some master plan that I myself was not aware of but was discovering as I went along.

The more I delved into evolutionary research, the more I got the feeling that the descriptions of evolution as violent and cruel were tendentious and systematic. What began to bother me was that this approach sat too well with what I myself already thought about evolution—like a true believer who is shown signs of revelation and nods and says "Yes, that is what I always thought." In that moment, I started to worry. Since when is science supposed to be so comfortable? Since when does science coincide with our basic intuitions? The more I looked into the findings, the more I understood that it's almost too easy, too comfortable to describe human nature in this way.

When I first heard how Yuval Noah Harari describes the disappearance of the Neanderthals, suddenly everything became as clear as day. Humans are cruel, violent, and bloodthirsty. That is our special skill; that is why we are alone. These explanations are convenient, too convenient. When I asked myself why they work, I understood that the main reason is that they allow us not only to explain but also and perhaps above all to excuse the apocalypse of the twentieth century. I was reminded of Primo Levi, protagonist of another book I wrote, Primo Levi and Ka-Tzetnik: *The Map and the Territory* (2021), and his need to explain Auschwitz, and I

[1] See: (Ataria et al., 2016).
[2] See: (Freud, 1939).

understood that for rational individuals who consider themselves members of Western culture and do not see that as a curse, Auschwitz is a phenomenon that demands an explanation. Auschwitz cannot simply be glossed over or ignored.

Auschwitz is what unites the various scientific disciplines: evolutionary, anthropological, and historical research all geared towards explaining the absolutely horrifying phenomenon of an enlightened people devising and embarking on an extermination project. Another phenomenon that demands explanation is the atomic bomb—and more than the bomb dropped on Hiroshima, the one dropped on Nagasaki, without any prior discussion. The automatism, which is very reminiscent of divine automatism in the book of Genesis (every time someone lays a hand on Sarah, for example—particularly in the case of Abimelech) cannot but horrify us. Why was the second bomb necessary, and if it was necessary, why did they not wait to see whether the Japanese would surrender?

* * *

The atomic bomb was an event in which Western culture was reborn—first, because it ended the Second World War, marking the beginning of a new era in which everything we previously knew about death and war is no longer relevant; second, after the bomb, every person in the street understands that humanity now has the power to destroy itself with disturbing ease; third, the intensity of the event, the shock, the speed, the density, the similarity to Western cultural ideas of divine rage (in the story of the destruction of Sodom, for example). All of these things immediately transformed the event into a myth.

While Auschwitz, a modern extermination operation, undermined the foundations of the Enlightenment, the atomic bomb instantly transformed the scientific story from wonder to horror. Oppenheimer and most of the other scientists working on the nuclear project understood this a moment too late. Techno-science has a dynamic of its own that appears to be unstoppable by anyone or anything. Scientific curiosity combined with the basic human aspiration to power, strength, and the feeling that if I don't do it someone else will, ensures that techno-science will continue to produce technologies of self-destruction. What is frightening is that there may be something deeper here, inherent to the scientific project, some self-destructive tendency of the techno-scientific project itself—perhaps as an ultra-religious and essentially apocalyptic project, as I suggested in *The Uncanny and Technology* (2023).

This insight sent me back to reread the various studies on the subject, and I quickly began to understand that Auschwitz is the collective unconscious that shapes Western thought in the present era. It is what unites research and researchers; it is what we all agree on. To understand the origin of humankind is to try to explain Auschwitz. All the rivers go to the sea, all the trains go to Auschwitz, and human history must also end at Auschwitz and Hiroshima.

Hiroshima does not allow us to think about the tool use of prehistoric *Homo*—whether *habilis*, *erectus*, or *sapiens*—without thinking about the bomb. It goes without saying that there is a gap between the use of flints and bones and dropping an atomic bomb. And yet it seems that Kubrick's scene grasps an unconscious truth: the moment we pick up tools everything changes.

The film *2001: A Space Odyssey* was released in 1968. Four years earlier, in 1964, *Dr. Strangelove* showed how easy it is to precipitate a nuclear war based on pure rational thought. In 1971, 3 years after 2001, Kubrick released *A Clockwork Orange*, which describes utterly nihilistic life in a post-nuclear reality. In one of the film's most famous scenes, the protagonist Alex murders a woman with a giant phallus, symbolizing violence, masculinity, and tools. This is the only way in which we manage to think about human history. Between *Dr. Strangelove* and *A Clockwork Orange*, Kubrick turned his attention to the myth of the first human, the first use of a bone that in a single gesture becomes perfect artificial intelligence that destroys humanity. *It is a male ape, male astronauts, and male technology. The synthesis of masculinity and technology ends in destruction.* This seems to be one of the clearest messages conveyed by Kubrick, perhaps the greatest mythographer since Homer.

* * *

I can divide my life into before Heidegger and after Heidegger. I came to Heidegger after a long journey with Husserl and Merleau-Ponty. I came hostile. I began to read and I felt that finally someone is organizing my perception of reality. He won me over. You don't see the world in the same way after Heidegger. You feel illuminated and exalted, moving through the world as if you are the only one who actually understands it. It is pleasant and dangerous. *Heidegger's attraction is the attraction of Nazism, or perhaps the opposite: the attraction of Nazism is the attraction of Heidegger.* This is perhaps the most important insight I have had in the course of writing this book. It is no coincidence that the Nazi movement managed to sweep Germany. The reasons are not only historical (the economic crisis and so forth). There is something attractive in the combination of power, kitsch, and death—a topic Friedländer addressed in his *Reflections on Nazism: An Essay on Kitsch and Death* (1984), which attempts to explain the allure of Nazism. In an era after the death of God, an era in which we are left merely with the shadow of God, the Nazis created something very close to a faith community, in which the masses feel that they are part of something big, that life is not one-dimensional. More than kitsch and death, however, it is the apocalypse.

In March 1945, when it was clear that Germany was headed for defeat, Hitler issued the Nero Decree, ordering the destruction of every area from which German forces withdrew. Essentially, Hitler asked Germans to destroy Germany. This decree reveals a deep truth about the Nazi movement: the desire for self-destruction. If the Nazi movement embodies humankind's deepest fall into technological being, then we understand that alongside the will to absolute and total power lies the desire for self-destruction and that the two elements are, in fact, different sides of the same coin.

Let us now return to Heidegger and his fundamental understanding that there was a deep link between his philosophy and the Nazi movement. These moments, even if they were just moments (which can be limited to 1930–1934, at most) should be taken utterly seriously. Heidegger identified in technology the will to power on the one hand and the tendency to self-destruction on the other—the same vector, the same trajectory. The very same tendency can be found in his philosophy, hence his deep connection to Nazism.[3]

At this point it is important to remember the observation of French philosopher Bernard Stiegler, who was, to my mind at least, the most important interpreter of Heidegger on the subject of technology and the most original and interesting thinker in this field: Heidegger was unable to think outside myth.[4] Heidegger diagnosed life in the technological era so well precisely because he himself was completely trapped in the greatest myth to shape Western thought: the apocalypse. Heidegger, I would argue, *was more of a prophet than a philosopher*, influenced far more by the book of Revelations than by philosophers such as Plato, Kant, and Schelling.

[3] אפשר בשלב זה להציע שהיידגר ניתח היטב את ההוויה הטכנולוגית רק שהיא לא ההוויה היהודית אלא ההוויה הנאצית (ובהקשר הזה מעניין לחשוב על היהודים ועל הגרמנים במונחים של כפילים)

[4] See: (Stiegler, 1998).

Heidegger, more than any other child of Western philosophy was a child of the Christian world view, perhaps, in his delusions of grandeur, seeing himself as the harbinger of the final book of the New Testament, that is, the one that will come after Revelations. Hence his awaiting the gods, hence his attraction—the shadow of a prophet in an age without prophecy. Heidegger, like many others in Western culture, awaits the return of Christ. In this context, perhaps the annihilation of the Jewish people is actually a good thing; perhaps that is why he never apologized. And perhaps it is better that way. Better silence than the hypocrisy of apology.

References

Ataria, Y. (2017). *The structural trauma of western culture: Toward the end of humanity*. Palgrave Macmillan.

Ataria, Y. (2021). *Primo Levi and Ka-Tsetnik*. Palgrave Macmillan.

Ataria, Y. (2023). *The uncanny and technology*. Pardes. (Heb.).

Ataria, Y., Gurevitch, D., Pedya, H., & Neria, Y. (Eds.). (2016). *Interdisciplinary handbook of culture and trauma*. Springer International.

Freud, S. (1939). *Moses and monotheism*. (K. Jones, Trans.). Vintage Books.

Friedländer, S. (1984). *Reflections of Nazism: An essay on kitsch and death*. Trans. Thomas Weyr. Harper & Row.

Stiegler, B. (1998). *Technics and time, 1: The fault of Epimetheus*. Stanford University Press.

Appendix: Autobiography of the Last Human

It never ends—all day, every day, 24/7/365. "It's the nightmare of having to present to the world from morning till night as a superman, our universal petty ideal, the grovelling sub-man we really are" (Céline, 1966, p. 361). Not a moment's rest: "Once one's in it, one's in it up to the neck" (p. 10). And then it's all one long decline: "You don't climb upwards in life; you go down" (p. 402). Enough! I can't take it any more: "I just want it over and done with as soon as possible!" (Musil, 1996, p. 260).

At this point in human history, with everything that we already know, there is no other way of putting it: "It is of men, and of them only, that one should always be frightened" (Céline, 1966, p. 14). Therefore, also, "To trust in men is itself to let oneself be killed a little" (p. 156). Indeed, "Anyone who wants to build on rock in dealing with human beings has to rely on the baser qualities and passions, for only what is most closely bound up with egotism endures and can always be counted on" (Musil, 1996, p. 553). It is clear to me that you already know what I am about to tell you, but I can't help myself. I am enjoying it. "I am a sick man.... I am a spiteful man. I am a most unpleasant man" and, to be honest, "It's out of spite that I don't wish to be treated" (Dostoevsky, 1989, p. 3). I'm not concerned that someone may be shocked. On the contrary, "It's extremely pleasant to hear such comments in our negative age" (ibid., p. 14). I won't give in this time to the art of concision: "Now I would like to tell you, gentlemen, whether or not you want to hear it, why it is that I couldn't

Y. Ataria, *Genes, Technology, and Apocalypse*,
https://doi.org/10.1007/978-3-031-59197-6

even become an insect" (ibid., p. 5). Of course, "I was monstrously exaggerating all these facts. But what could be done? I was no longer able to control myself" (ibid., p. 47). Indeed, one can sink even lower and there is no escape from the underground: "I indulged in depravity all alone, at night, furtively, timidly, sordidly, with a feeling of shame that never left me even in my most loathsome moments and drove me at such times to the point of profanity. Even then I was carrying around the underground in my soul" (ibid., p. 33). But it is a pleasure:

> I used to gnaw and gnaw at myself inwardly, secretly, nagging away, consuming myself until finally the bitterness turned into some kind of shameful, accursed sweetness and at last into genuine, earnest pleasure! Yes, into pleasure, real pleasure! ... But it's precisely in that cold, abominable state of half-despair and half-belief, in that conscious burial of itself alive in the underground for forty years because of its pain, in that powerfully created, yet partly dubious hopelessness of its own predicament, in all that venom of unfulfilled desire turned inward, in all that fever of vacillation, of resolutions adopted once and for all and followed a moment later by repentance—herein precisely lies the essence of that strange enjoyment I was talking about earlier. (ibid., pp. 6–9)

Céline—who, as a French pacifist and patriot during the Second World War, was concerned by the slow pace at which the Jews were being exterminated—has his protagonist Ferdinand begin life "with such enthusiasm... I admit that. ... Enthusiasm, alas, was... reserved for us" (1966, pp. 24–33). There is a cure, however, for this disease ... How shall I put it? Well ... life itself: "Electrical treatment of the body and strenuous doses of the ethics of patriotism for the soul" (ibid., p. 83). Electric shocks are great for learning a little love of country, but if you really want to know what love is, there's nothing like a sweet little war:

> You felt your way ... staggering ... from one unknown barn to another, shouted at, threatened, haggard and without hope of any end to it all other than in slush amid oaths and in disgust at having been tortured and duped to death by a horde of vicious madmen, who had suddenly become incapable of doing anything else as long as they lived, but kill and be slit in half without knowing the reason why. (ibid., p. 30)

We love to give ourselves credit at least when it comes to murder—something to show for all that progress—but let's be honest, it's not that we're particularly good at it, but, over the years, we have managed to sink a little lower: "If man hasn't become more bloodthirsty as a result of civilization, surely he's become bloodthirsty in a nastier, more repulsive way than before" (Dostoevsky, 1989, p. 17), and, altogether, "Since there's no sense in our life as a whole, neither is there such a thing as progress as a whole" (Musil, 1996, p. 528).

People generally relate to youth as a glorious, magical time of life, but what is youth if not the age at which you turn into a blob of hormones that make you venerate women and give you erections, pimples, and the belief that there is a way out, that things can be different, God help us. It is only a lack of consciousness that allows us not to kill ourselves at this charming age. When our consciousness finally begins to develop, we must admit that it was, in fact, one of the lowest times of our lives: "You are already rather less proud of this youth of yours, you don't yet dare to announce [it] in public" (Céline, 1966, p. 250). Of course, one can always argue that "being overly conscious is a disease" (Dostoevsky, 1989, p. 5). Nevertheless, I'm not sure we would be willing to give it up: "Suffering is the sole cause of consciousness. Although I stated earlier that in my opinion consciousness is man's greatest misfortune, still I know that man loves it and would not exchange it for any other sort of satisfaction" (ibid., p. 24).

The glory days of youth are the time at which wretchedness, initially well-disguised, begins to take shape. A splitting pain, drained only by masturbation—which, in adolescence, seems like a possible cure... well... for life itself. Growing up is the understanding that life cannot be based solely on masturbation, or, in the words of Louis C.K.: "I need to cum, I need to—cumming is a need, I came the first time when I was 12 and I haven't skipped a day." A good education helps: "She hadn't enough education to follow me now ... no education ... We'd understood each other well enough at one time, the Henrouille woman and I ... Quite a while that had lasted ... But now she wasn't low enough down for me, she couldn't get down ... Didn't know how to join me ... She hadn't the brains or the strength" (Céline, 1966, p. 402). Indeed, at a certain point, weariness is all that remains: "The great weariness of life is maybe nothing but the vast trouble we take to remain always for twenty or forty or more years at a time reasonable beings—so as not to be merely and profoundly oneself, that is to say, obscene, ghastly, and absurd" (ibid., p. 361). The more honest we are with ourselves, the more we remove our masks and allow ourselves to sink, the more alone we find ourselves.

Pimpled youth, when you sell yourself cheaply—but even at bargain-basement prices, there are no takers and indeed "You're of no interest to them, the swine, except when you're bleeding" (ibid., p. 71)—becomes a journey on which you discover just how low you can sink, although it is frankly surprising how many more floors there are below:

> When one is still young, one manages to excuse the stoniest indifferences, the most bare-faced impositions, making allowances for them on grounds of personal idiosyncrasy and God knows what sort of callow romanticism. But later on, when life has taught you how much caution, cruelty and malice are required of you, if only you are to live through it reasonably well at a normal temperature, you realize, you're in a position to understand the dirty deals a past contains. All that you need to do, and what you always need, is to take a very careful look into your own heart and see what you yourself have sunk to. There's no mystery, no nonsense left; you've had to swallow all the poetry you ever had, since you've survived till then. Life's become a lot of pork and beans. (ibid., p. 184)

At this point, it really is not clear "why people don't dare to kill themselves" (Dostoyevsky, 1994, p. 113). The answer to this question, according to Dostoyevsky is pain and prejudice. And despite the fact that it is fear of pain that prevents mass suicide, pain ultimately encroaches on every aspect of life:

> And between whiles you boast of having managed to be rid of your unhappiness, but everyone knows, don't they, that you've not done any such thing but are just as miserable as you well can be. As you grow uglier and uglier and more repulsive, playing this game as you grow old, you can't even conceal your unhappiness any longer, your failure; and in the end your face has become only an ugly expression which takes twenty, thirty or more years to come up from your stomach on to your face. That is what a man will achieve, that and that alone—an ugly expression which he has spent a lifetime making and often even then hasn't managed to finish off properly, because of how difficult and complicated an expression it would have to be to reflect his real soul without missing anything out. (Céline, 1966, p. 254–255).

Yet, despite it all (books, films, restaurants, shopping, flights, sports, etc.), despite Darwin and Nietzsche (who must be mentioned if one is ever to get a professorship), despite having solved the problem of how to organize the various books on the shelves (some of which belong in different categories)—another wretched and beloved cliché of wannabe nihilistic intellectuals—at the very end of the road, we seem, for some reason, to need people:

> He was crying. He himself had come to the end of things. There wasn't anything you could tell him now. There is a moment when you are all alone and by yourself and have come to the end of all that can happen to you. It's the end of the world. Unhappiness itself, your own misery, won't answer you now and you have to go back, among men, no matter where. One isn't difficult at moments like that, for even to weep you've got to get back to where everything starts, to where the others are. (ibid., p. 285)

But the need for people and companionship is not about love; it's about the need to be a despot: "As the years went by, my need for people, for friends, increased. … Once I even had a friend of sorts. But I was already a despot at heart; I wanted to exercise unlimited power over his soul…. It was the power I needed then, I craved the sport, I wanted to reduce you to tears, humiliation, hysteria" (Dostoevsky, 1989, pp. 47–83). There's nothing like fantasies of total domination. Thus, when Ulrich, *The Man without Qualities*, is asked what he would do if he could rule the world for a day, his answer, unlike his usual prevarications throughout the book, is rather simple: "I suppose I would have no choice but to abolish reality" (Musil, 1996, p. 312). That would indeed appear to be the necessary outcome, since "Historic periods of permissiveness have never failed, so far, in making all those who lived in them miserable" (ibid., p. 548).

There is no more hackneyed product than love. Yet, sales are booming. But perhaps when we still need someone, we have not yet sunk low enough: "If you really want to know the whole of it... why, it's every darn thing that repels me and disgusts me now. Not only you! Everything! … Love especially" (Céline, 1966, p. 430).

You have to accept that ultimately it's all matter, and believe in evolution too. But in everyday life, this evolution is a curse. Our organism is too talented to survive: "When one's in this world, surely the best thing one can do, isn't it, is to get out of it?" (ibid., p. 53). But this evolution is stuck to our bones, which is why we can be pushed further and further down. Evolution, the great Theory sanctified by science, is a huge joke at our expense: "The bodies we possess, a fancy dress of twitching, trivial molecules, *revolt unceasingly against this frightful farce of managing to last.* They want to be off, these molecules of ours, and lose themselves, as quick as they can, in the universe at large: little beauties! They hate just being 'us', mere *cuckolds of the Infinite*!" (ibid., p. 293). It's not exactly clear why: "If the laws of nature did not pity even This One, did not pity even their own miracle, but made Him, too, live amidst a lie and die for a lie, then the whole planet is a lie, and stands upon a lie and a stupid mockery. Then the very laws of the planet are a lie and a devil's vaudeville" (Dostoyevsky, 1994, p. 618).

Evolutionary programming glues us to life, which is, needless to say, incredibly boring: "Of course, a moment or two later I would realize in anger that it was all lies, lies, revolting made-up lies, that is, all that repentance, all that tenderness, all those vows to mend my ways. But you'll ask why I mauled and tortured myself in that way? The answer is because it was so very boring" (ibid., p. 12). Boredom is a threat, because it reminds us of our fate, since only boredom allows us to sense time: "A day that is nothing but a mere round of the twenty-four hours isn't to be borne" (Céline, 1966, p. 307). But time, and here's the beauty of the thing, doesn't simply destroy us; it torments us, allows our bodies to be devoured, until our spines need to be stapled together, lest they crumble. Entertainment and more entertainment is the answer: "Artistes are stuck about everywhere these days, as a safety measure, with a view to our not being too bored. You have them all over every building now, overflowing with thrills, their passionate revelations trickling from floor to floor" and "Nowadays public lavatories go in for décor, and so does the slaughterhouse and the pawnshop too—just to keep you amused and distract you, and let you forget your fate" (ibid., p. 306). A motto like this is just the thing to get us to build an atomic bomb, and, sure enough, "Oppie encouraged everyone to work hard and play hard" (Bird & Sherwin, 2005, p. 257).

We must therefore continue our descent, further and further down. "We'd all get somewhere together in the end and then we'd find out what it was we'd been looking for in this whole business. That is what life boils down to, a light going out in the darkness." If the journey is successful, then there is a real reason to be proud: "The people who are out in the daylight can't understand you now" (Céline, 1966, p. 295). The light has gone out, night has fallen, you feel at home.

References

Agamben, G. (2002). *Remnants of Auschwitz: The witness and the archive. Homo Sacer III* (D. Heller-Roazen, Trans.). Zone Books.

Ardrey, R. (1976). *The hunting hypothesis: A personal conclusion concerning the evolutionary nature of man.* Atheneum.

Arendt, H. (1979). *The origins of totalitarianism.* Harcourt Brace and Company.

Arendt, H. (2006). *Eichmann in Jerusalem: A Report on the banality of evil.* Penguin Books.

Arendt, H., & Heidegger, M. (2004). *Letters 1925–1975* (U. Ludz, Ed., & A. Shields, Trans.). Harcourt.

Arendt, H., & Jaspers, K. (1992). *Hannah Arendt/Karl Jaspers Correspondence, 1926–1969* (L. Kohler, H. Saner, Eds., R. Kimber, & R. Kimber, Trans.). Harcourt Brace & Company.

Ataria, Y. (2024). The uncanny and technology. In *In an effort to imitate the act of creation (Heb.).* Pardes.

Bacon, F. (2020). In G. B. Wegemer (Ed.), *New Atlantis.* CTMS Publishers at the University of Dallas.

Baggott, J. (2010). *The first war of physics: The secret History of the atom bomb 1939–1949.* Pegasus Books.

Bates, T. B. (2016, March 9). Early human habitat, recreated for first time, shows life was no picnic. *Rutgers Today.* https://www.rutgers.edu/news/early-human-habitat-recreated-first-time-shows-life-was-no-picnic

Baudrillard, J. (2004). *Simulacra and simulation* (S. F. Glaser, Trans.). University of Michigan Press.

Baynes, N. H. (Ed.). (1942). *The Speeches of Adolf Hitler, April 1922–August 1939* (Vol. 1) (B. N. H, Trans.). Oxford University Press.

Bird, K., & Sherwin, M. J. (2005). *American Prometheus: The triumph and tragedy of J. Robert Oppenheimer.* Atlantic Books.

Bloch, E. (1970a). *A philosophy of the future* (J. Cumming, Trans.). Herder and Herder.

Bloch, E. (1970b). *Tübinger Einleitung in die Philosophie, Gesamtausgabe* (Vol. 13). Suhrkamp.

Bretas, R., Yamazaki, Y., & Iriki, A. (2020). Phase transitions of brain evolution that produced human language and beyond. *Neuroscience Research, 161*, 1–7. https://doi.org/10.1016/j.neures.2019.11.010

Buber, M. (1958). *Paths in Utopia* (R. F. Hull, Trans.). Beacon Press.

Bunn, H. T. (2007). Meat made us human. In P. S. Ungar (Ed.), *Evolution of the human diet: The known, the unknown, and the unknowable* (pp. 191–211). Oxford University Press.

Burdett, M. S. (2014). *Eschatology and the technological future.* Routledge.

Cartmill, M. (1993). *A view to a death in the morning: Hunting and nature through history*. Harvard University Press.

Céline, L.-F. (1966). *The journey to the end of the night*. Penguin Books.

Chomsky, N. (2016). *What kind of creatures are we?* Columbia University Press.

Claeys, G. (2017). *Dystopia: A natural history: A study of modern despotism, its antecedents, and its literary diffractions*. Oxford University Press.

Clark, T. (2011). *Martin Heidegger*. Routledge.

Cohen, A. (1989). *Nuclear age: A chapter in moral history (Heb.)*. Ministry of Defence.

Compton, A. H. (1956). *Atomic quest: A personal narrative*. Oxford University Press.

Dan, J. (2000). *Apocalypse then and now (Heb.)*. Sifre Hemed.

Dan, J. (2009). *Chaos theory and the study of history (Heb.)*. Kinneret, Zmora-Bitan, Dvir.

Dart, R. (1925). Australopithecus africanus the man-ape of South Africa. *Nature, 115*, 195–199.

Dart, R. (1953). The predatory transition from ape to man. *International Anthropological and Linguistic Review, 1*(4), 201–217.

Darwin, C. (1998). In G. Beer (Ed.), *The origin of species*. Oxford University Press.

Dawkins, R. (2006). *The selfish gene*. Oxford University Press.

de Waal, F. (2005). *Our inner ape: A leading primatologist explains why we are who we are*. Riverhead Books.

Derrida, J. (1989). *On Spirit: Heidegger and the Question* (G. Bennington, & R. Bowlby, Trans.). University of Chicago Press.

Di Cesare, D. (2018). *Heidegger and the Jews: The Black Notebooks* (M. Baca, Trans.). Polity Press.

Diamond, J. (2006). *The third chimpanzee: The evolution and future of the human animal*. Harper Perennial.

Diamond, J. (2012). *The world until yesterday: What can we learn from traditional societies?* Viking.

Dostoevsky, F. (1989). *Notes from Underground* (M. R. Katz, Trans.). W. W. Norton and Company.

Dostoyevsky, F. (1994). *Demons* (R. Pevear, & L. Volokhonsky, Trans.). Alfred A. Knopf.

Dunsworth, H. M. (2010). Origin of the genus homo. *Evolution: Education and Outreach, 3*, 353–366.

Eakin, E. (2002, May 18). Before the word, perhaps the wink?; Some language experts think humans spoke first with gestures. *The New York Times*.

Eatherly, C., & Anders, G. (1962). *Burning conscience: The case of the Hiroshima pilot, Claude Eatherly, told in his letters to Günther Anders*. Monthly Review Press.

Eliot, T. S. (1972). *Four quartets*. Faber and Faber.

Ellul, J. (1964). *The technological society* (J. Wilkinson, Trans.). Vintage Books.

Ellul, J. (1980). *The technological system* (J. Neugroschel, Trans.). Wipf and Stock Publishers.

Eshchar, Y. (2022, February 21). *When did we come to Europe: Earlier than we thought (Heb.)*. Retrieved from Davidson Institute of Science Education: Science News: https://davidson.weizmann.ac.il/online/sciencenews/

Felman, S., & Laub, D. (1992). *Testimony: Crises of witnessing in literature, Psychoanalysis and history.* Routledge.

Ferguson, B. (2013). Pinker's list. In D. P. Fry (Ed.), *War, peace, and human nature: The convergence of evolutionary and cultural views* (pp. 112–131). Oxford University Press.

Finlayson, C. (2009). *The humans who went extinct: Why neanderthals died out and we survived.* Oxford University Press.

Foucault, M. (1984). What is enlightenment? In M. Foucault, & P. Rabinow (Ed.), *The Foucault Reader* (C. Porter, Trans. pp. 32–50). Pantheon Books.

Foucault, M. (1995). *Discipline and punish: The birth of the prison* (A. Sheridan, Trans.). Vintage Books.

Foucault, M. (2005). *The order of things: An archaeology of the human sciences.* Routledge.

Fraenkel, D. (1999). The abnormality of the normal. *Yad Vashem Studies, 27*, 429–439.

Friedländer, S. (1997, 2007). *Nazi Germany and the Jews.* HarperCollins.

Fry, D. P. (2013). War, peace, and human nature. In D. P. Fry (Ed.), *War, peace, and human nature* (pp. 1–21). Oxford University Press.

Gat, A. (2006). *War in human civilization.* Oxford University Press.

Ghiglieri, M. P. (1999). *The dark side of man: Tracing the origins of male violence.* Helix.

Gissis, S. (2020, May). The Racial assumption in genomic science (Heb.). *Hazman Hazeh.*

Goldstein, J. S. (2011). *Winning the war on war: The decline of armed conflict worldwide.* Penguin.

Graeber, D., & Wengrow, D. (2021). *The Dawn of everything: A new history of humanity.* Penguin.

Graur, M., & Graur, D. (1992). *Eugenics: Improving the hereditary quality of man (Heb.).* Ministry of Defence.

Greene, B. (2004). *The fabric of the cosmos: Space, time, and the texture of reality.* Alfred A. Knopf.

Habermas, J. (1998). Martin Heidegger: On the publication of lectures from the year 1935. In R. Wolin (Ed.), *The Heidegger controversy: A critical reader* (pp. 186–197). MIT Press.

Habermas, J. (2003). *The future of human nature* (W. Rehg, M. Pensky, & H. Beister, Trans.). Polity.

Hamilton, W. D. (1996). *Narrow roads of gene land* (Vol. 1). W. H. Freeman.

Harari, Y. N. (2014). *Sapiens: A brief history of humankind.* Vintage Books.

Harari, Y. N. (2016). *Homo Deus*. Random House.
Harkaby, Y. (1990). *War and strategy (Hebrew)*. Ma'arachot.
Harman, O. (2011). *The Price of altruism: George Price and the search for the origins of kindness*. W. W. Norton & Company.
Hass, R. (2020). Female hunters of the early Americas. *Scientific Advances, 6*(45), eabd0310. https://doi.org/10.1126/sciadv.abd0310
Heidegger, M. (1949). Hölderlin and the essence of poetry. In M. Heidegger (Ed.), *Existence and being* (pp. 291–315). Vision.
Heidegger, M. (1966). *Discourse on thinking* (J. M. Anderson, & E. H. Freund, Trans.). Harper and Row.
Heidegger, M. (1973). Overcoming metaphysics. In M. Heidegger (Ed.), *The End of Philosophy* (J. Stambaugh, Trans. pp. 84–110). Harper and Row.
Heidegger, M. (1977). *The question concerning technology and other essays* (W. Lovitt, Trans.). Harper & Row.
Heidegger, M. (1981). "only a god can save us": The Spiegel interview (1966). In T. Sheehan (Ed.), *Heidegger: The Man and the Thinker* (W. J. Richardson, Trans., pp. 45–67). Precedent Publishing.
Heidegger, M. (2000). *Introduction to metaphysics* (G. Fried, & R. Polt, Trans.). Yale University Press.
Heidegger, M. (2003). The self-assertion of the German university. In M. Heidegger, & M. Stassen (Ed.), *Philosophical and political writings* (K. Harries, Trans., pp. 2–11). Continuum.
Heidegger, M. (2006). In W. McNeill (Ed.), *Pathmarks*. Cambridge University Press.
Heidegger, M. (2008a). The origin of the work of art. In M. Heidegger & D. F. Krell (Eds.), *Basic writings: From being and time (1927) to the task of thinking (1964)* (pp. 139–212). Harper Perennial.
Heidegger, M. (2008b). Letter on humanism. In M. Heidegger & D. F. Krell (Eds.), *Basic writings: From being and time (1927) to the task of thinking (1964)* (pp. 213–265). Harper Perennial.
Heidegger, M. (2008c). On the essence of truth. In M. Heidegger & D. F. Krell (Eds.), *Basic writings: From being and time (1927) to the task of thinking (1964)* (pp. 111–138). Harper Perenniel.
Heidegger, M. (2012). *Bremen and freiburg lectures* (A. J. Mitchell, Trans.). Indiana University Press.
Heidegger, M. (2017). *Ponderings XII–XV: Black notebooks 1939–1941* (R. Rojcewicz, Trans.). Indiana University Press.
Heidegger, M., & Marcuse, H. (2004). Heidegger and Marcuse: A dialogue in letters. In H. Marcuse, & D. Kellner (Ed.), *Collected Papers of Herbert Marcuse, Vol. 1: Technology, War and Fascism* (W. Richard, Trans., pp. 261–267). Routledge.

Hershkovitz, I., May, H., Sarig, R., et al. (2021). A middle Pleistocene homo from Nesher Ramla, Israel. *Science, 372*(6549), 1424–1428. https://doi.org/10.1126/science.abh3169

Horkheimer, M. (1978). *Dawn and Decline: Notes 1926–1931 and 1950–1969* (M. Shaw, Trans.). The Seabury Press.

Horkheimer, M. (2004). *Eclipse of reason.* Continuum.

Horkheimer, M., & Adorno, T. W. (2002). In G. S. Noerr (Ed.), *Dialectic of Enlightenment* (E. Jephcott, Trans.). Stanford University Press.

Huxley, A. (1968). *Brave New World.* Bantam Books.

Huxley, A. (1969). In G. Smith (Ed.), *Letters of Aldous Huxley.* Chatto & Windus.

Ihde, D. (2010). *Heidegger's technologies.* Fordham University Press.

Jablonka, E., & Lamb, M. J. (2005). *Evolution in four dimensions: Genetic, epigenetic, Behavioral, and symbolic variation in the history of life.* MIT Press.

Jonas, H. (1984). The imperative of responsibility. In *Search of an Ethics for the Technological Age* (H. Jonas, & D. Herr, Trans.) Chicago University Press.

Kravitz, A. (2019). Heidegger the Nazi, Heidegger the antisemite—another look (Heb.). *Tabur, 9*, 23–27.

Kubrick, S. (Director). (1964). *Dr. Strangelove or: How I Learned to Stop Worrying and Love the Bomb* [Motion Picture].

Kubrick, S. (Director). (1968). *2001: A space odyssey* [Motion Picture].

LaCapra, D. (2014). *Writing history, writing trauma.* Johns Hopkins University Press.

Lapidot, E. (2020). Is technology Jewish? A conversation with Heidegger. *Journal of Jewish Thought and Philosophy, 28*, 95–120.

Latour, B. (1994). On technical mediation: Philosophy, sociology, genealogy. *Common Knowledge, 3*(2), 29–64.

Leakey, R. E. (1994). *The origin of humankind.* London.

Leakey, R. E., & Lewin, R. (1979). *People of the Lake: Mankind and its beginnings.* Avon.

Lemmerich, J. (2011). *Science and conscience: The life of James Franck* (H. A. M, Trans.). Stanford University Press.

Leroi-Gourhan, A. (1993). Gesture and speech (A. B. Berger, Trans.). MIT Press.

Levinas, E. (1989). As if consenting to horror. *Critical Inquiry, 15*(2), 485–488.

Lévi-Strauss, C. (1952). *Race and history.* UNESCO.

Lévi-Strauss, C. (1971). Race and culture. *International Social Science Journal, XXIII*(4), 608–625.

Lewis, J. E., & Harmand, S. (2016). An earlier origin for stone tool making: Implications for cognitive evolution and the transition to homo. *Philosophical Transactions of the Royal Society B: Biological Sciences, 371*, 20150233. https://doi.org/10.1098/rstb.2015.0233

Löwith, K. (1993). The political implications of Heidegger's existentialism. In R. Wolin (Ed.), *The Heidegger controversy: A critical reader* (pp. 167–184). MIT Press.

Lyotard, J. F. (1990). *Heidegger and "the Jews"*. (A. Michel, & M. S. Roberts, Trans.). University of Minnesota Press.

Mansbach, A. (1998). *Existence and meaning (Heb.)*. Magnes.

Marcuse, H. (1968). *One dimensional man: Studies in the ideology of advanced industrial society*. Beacon Press.

Marx, K. (1955). *The poverty of philosophy*. Progress Publishers.

Milo, D. S. (2009). *The invention of tomorrow (Heb.)*. Hakibbutz Hameuchad.

Musil, R. (1996). *The man without qualities: A sort of introduction and pseudoreality prevails* (S. Wilkins, Trans.). Vintage International.

Nancy, J. L. (2017). *The banality of heidegger* (J. Fort, Trans.). Fordham University Press.

Nietzsche, F. (1974). *The gay science* (W. Kaufmann, Trans.). Vintage Books.

Nietzsche, F. (1996). *Human, all too human*. Cambridge University Press.

Ophir, A. (2005). *The Order of evils: Toward an ontology of morals* (R. Mazali, & H. Carel, Trans.). Zone Books.

Ott, H. (1994). *Martin Heidegger: A political life* (A. Blunden, Trans.). Fontana Press.

Pimentel, D. (2019). *Heidegger with Derrida: Being Written* (N. Olshansky-Ashtr, Trans.). Palgrave Macmillan.

Popper, K. R. (1963). *The open society and its enemies* (Vol. 2). Princeton University Press.

Ridley, M. (1996). *The origins of virtue*. Viking.

Ridley, M. (2003). *Nature via nurture: Genes, experience, and what makes us human*. Fourth Estate.

Roosevelt, G. G. (1987). A reconstruction of Rousseau's fragments on the state of war. *History of Political Thought, 8*(2), 225–244.

Rutherford, A. (2018). *A brief history of everyone who ever lived*. Weidenfeld & Nicolson.

Sagi, A. (2015). Tikkun Olam: Between utopian idea and socio-historical process. In A. Sagi, H. Tirosh-Samuelson, & A. W. Hughes (Eds.), *Existentialism, pluralism, and identity* (pp. 103–122). Brill.

Sartre, J.-P. (1956). *Being and nothingness: An essay on phenomenological ontology*. Philosophical Library.

Scholem, G. (1995). *The messianic idea in judaism and other essays on jewish spirituality* (M. A. Meyer, & H. Halkin, Trans.). Schocken Books.

Shapin, S. (1998). *The scientific revolution*. University of Chicago Press.

Sheehan, T. (1988). Heidegger and the nazis. *The New York Review of Books, 35*(10), 38–47.

Sluga, H. (1995). *Heidegger's crisis: Philosophy and politics in Nazi Germany*. Harvard University Press.

Speer, A. (1970). *Inside the third Reich* (R. Winston, & C. Winston, Trans.). Weidenfeld and Nicolson.

Speth, J. D. (2010). *The paleoanthropology and archaeology of big-game hunting: Protein, fat, or politics?* Springer.

Stangneth, B. (2016). *Eichmann before Jerusalem. The unexamined life of a mass murderer* (R. Martin, Trans.). Vintage.

Steiner, G. (1987). *Martin Heidegger*. University of Chicago Press.

Sussman, R. W. (1999). The myth of man the hunter, man the killer and the evolution of human morality. *Zygon, 34*(3), 453–471.

Sussman, R. W. (2013). Why the legend of the killer ape never dies: The enduring power of cultural beliefs to distort our view of human nature. In D. P. Fry (Ed.), *War, peace, and human nature: The convergence of evolutionary and cultural views* (pp. 97–111). Oxford University Press.

Thorpe, S. K., Holder, R. L., & Crompton, R. H. (2007, June 1). Origin of human bipedalism as an adaptation for locomotion on flexible branches. *Science, 316*(5829), 1328–1331. https://doi.org/10.1126/science.1140799

Trawny, P. (2016). Heidegger and the Shoah. In I. Farin & J. Malpas (Eds.), *Reading Heidegger's black notebooks 1931–1941* (pp. 169–179). MIT Press.

Truman, H. S. (1965). *Memoirs by Harry S. Truman* (Vol. 1). Signet Books.

USAEC. (1954). *In the matter of J. Robert Oppenheimer: Transcript of hearing before personnel security board, Washington D.C., April 12, 1954 through May 6, 1954.* United States Government Printing Office.

Vioulac, J. (2021). *Apocalypse of truth: Heideggerian meditations* (M. J. Peterson, Trans.). The University of Chicago Press.

Wolin, R. (2015). *Heidegger's children: Hannah Arendt, Karl Löwith, Hans Jonas, and Herbert Marcuse*. Princeton University Press.

Wrangham, R. (2009). *Catching fire: How cooking made us human*. Profile Books.

Wrangham, R. (2019). *The goodness paradox: The strange relationship between virtue and violence in human evolution*. Pantheon.

Wrangham, R., & Peterson, D. (1996). *Demonic males: Apes and the origins of human violence*. Mariner/Houghton Mifflin.

Wyatt, S. (2007). Technological determinism is dead. In E. J. Hackett, O. Amsterdamska, M. Lynch, & J. Wajcman (Eds.), *Handbook of science and technology studies* (pp. 165–180). MIT Press.

Zimmerman, M. E. (1990). *Heidegger's confrontation with modernity: Technology, politics, and art*. Indiana University Press.

Name Index[1]

[1] Note: Page numbers followed by 'n' refer to notes.

Y. Ataria, *Genes, Technology, and Apocalypse*,
https://doi.org/10.1007/978-3-031-59197-6

Subject Index[1]

[1] Note: Page numbers followed by 'n' refer to notes.

Y. Ataria, *Genes, Technology, and Apocalypse*,
https://doi.org/10.1007/978-3-031-59197-6

SPRINGER NATURE

GPSR Compliance

The European Union's (EU) General Product Safety Regulation (GPSR) is a set of rules that requires consumer products to be safe and our obligations to ensure this.

If you have any concerns about our products, you can contact us on ProductSafety@springernature.com

In case Publisher is established outside the EU, the EU authorized representative is:

Springer Nature Customer Service Center GmbH
Europaplatz 3
69115 Heidelberg, Germany

FSC
www.fsc.org
MIX
Papier | Fördert gute Waldnutzung
FSC® C083411

Kolibri 360 GmbH
Ferdinand-Jühlke-Straße 7
99095 Erfurt, Deutschland
produktsicherheit@kolibri360.de